U.S. Army Installations
World War II
A Research Reference

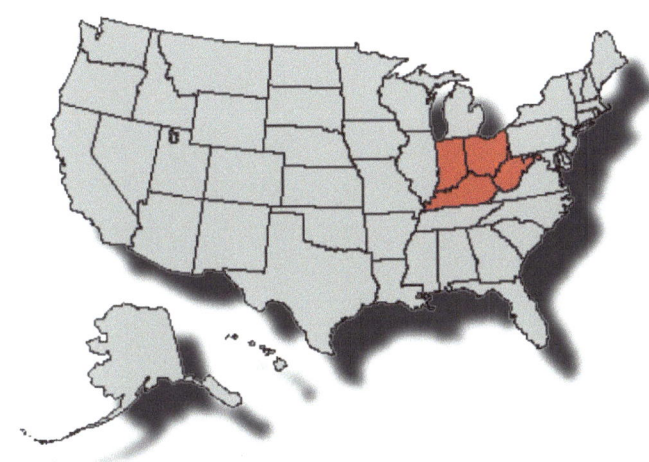

Fifth Service Command

Army Ground Forces
Army Air Forces
Army Service Forces
Prisoner of War Camps

Gerald Burke Leonard

Hollywood Beach Publishing

This document is not a publication of the United States Army,
the Department of Defense or the United States government in general

Copyright 2013 by Gerald B. Leonard
Library of Congress Control Number
2015904295

All rights reserved. No part of this book may be used or reproduced in
any manner whatsoever without written permission from the Publisher
except in the case of brief quotations embodied in critical articles or reviews.
For information address publisher at:

Hollywood Beach Publishing
2201 Marylyn Circle
Petaluma, CA 94954

ISBN 978-0-9891902-5-1

This paperback edition published in 2015

Second Printing - Remastered

Forward

Other authors have written about forts that have existed during the history of our country. Much has been written about the installations that were created for the Revolutionary War, as well as those that were built for the Civil War, the War of 1812, and the Indian Wars (Roberts). [1] This series of books deals only with the installations that existed or were created for the Second World War. Some has even been written about these facilities but not in the same context as that provided for by this series. In the case of the former, it is a reasonable accounting of the various forts that existed during the wars in the country's early history. In the case of Osborne, [2] the volume could largely be described as a travel guide. It is an excellent resource.

U.S. Army Facilities – World War II: A Research Reference approaches the study of forts and other such facilities that existed during that specific time period in a substantially different manner. First, is the identification of "all" the facilities that existed whereas Osborne excluded Alaska and Hawaii both of which had numerous WWII facilities. Second, the documentation is established to detail the actual facility in terms of its physical parameters. Third, and unprecedented, we have included a detailed accounting of the units that served at each facility. These data were drawn from Station Lists of the U.S. Army and the Directory of the U.S. Army. The data are documented by month for a given period of the conflict years and bimonthly for the balance. As this material does not exist in any electronic form, it has had to be photocopied and transcribed. Each of the 50 or so volumes is comprised of about 60,000 lines of data as entered into a database prior to incorporation into the text.

The facilities have been grouped according to the (sub) branch that they served. That is to say, if the Army is a branch of the service as is the U.S. Navy, then the Army Ground Forces, the Army Air Forces, the Army Service Forces are shown here as sub-branches of the U.S. Army. Prisoner of War camps have also been included in these data. They were operated under the jurisdiction of the Provost Marshal General and consisted of installations that in many cases comprised a portion of an army base. They were also established as independent facilities allied to a given community where farm labor was required of the war prisoners. Prisoners were also held at various army general hospitals, at army air fields and other such facilities.

The information that was collected to comprise these volumes came from a wide range of resources. A host of internet sites were invaluable in providing essential information. However, the authors also made site visits to a great many of the facilities. Some active bases refused our entry despite our benign purposes for the visit. However, we are in a period of conflict and base security is eminently more important than our access for research purposes. At locations where we were refused entry or where the facility has been razed, as well as all other locations we visited, we took advantage of interviews with whomever we could find with a direct connection to the base in question and always visited the local library. The information from local libraries always proved to be more extensive than that found on the internet or other generalized sources.

While our goal was to document 100% of all bases, posts, and other facilities used by the U.S. Army during WWII, needless to say such a goal is both imponderable and unachievable. We did our best and the reader will find a more extensive "picture" here of a base than would be the case in other such documentation.

From an organizational perspective we compiled our information by base, by type, i.e., AGF, AAF, ASF, etc. by state and by service command. CONUS, the Continental United States, was divided into 10 organizational domains called service commands – 9 service commands and the Military District of Washington. We have added an element beyond the CONUS perspective to pick up those forts, camps, etc. that were a critical component of our national defense posture. Generally these are identified as Defense Commands but also include Base Commands and Departments in the case of Alaska and Hawaii.

---

1   Roberts, Robert B., Encyclopedia of Historic Forts: *The Military, Pioneer, and Trading Posts of the United States*, MacMillan Publishing Company, New York, 1988.

2   Osborne, Richard E., *World War II Sites in the United States: A Tour Guide Directory*, Riebel-Roque Publishing Company, Indianapolis, IN, 1996.

To complement the reportage that these volumes represent, the authors have also created a database that embellishes all this information into a series of menu-driven components so that a reader can conduct queries related to any of the locations documented these volumes / database or by units that served at those facilities. Contact the publisher for details about the database program written in Microsoft Access and Microsoft Excel.

Table of Contents

Preface — i

Table of Contents — iii

Fifth Service Command Introduction — 1

Indiana — 5
    Army Ground Forces — 7
    Army Air Forces — 12
    Army Service Forces — 24
    Prisoner of War Camps — 41

Kentucky — 47
    Army Ground Forces — 49
    Army Air Forces — 56
    Army Service Forces — 64
    Prisoner of War Camps — 71

Ohio — 75
    Army Ground Forces — 77
    Army Air Forces — 81
    Army Service Forces — 88
    Prisoner of War Camps — 95

West Virginia — 99
    Army Ground Forces — 101
    Army Air Forces — 102
    Army Service Forces — 103
    Prisoner of War Camps — 107

Appendices — 105
    Army Ground Force Units — 107
    Army Air Force Units — 151
    Army Service Force Units — 179
    Prisoner of War Camp Units — 193
    Acknowledgements — 201
    Terminology & Bibliography — 203

Fifth Service Command

Introduction

Each volume of this series begins with the same orientation to the documentation of the research. While duplicatory, there are those who may elect to examine only one volume and they would require the same background as the individual who studies two or more of these volumes. Therefore, for those interested in more than just this one specific Service Command, the information in this Introduction will be found to be repeated from each other volume.

The information from this research has been organized by operational branch of the U.S. Army (from the World War II era); it has been grouped according to each of the Service Commands (geographically) and subsequently by state within each of the service commands. The data represent only the CONUS designation - the Continental United States. There were numerous bases, posts and other stations that formed a defensive ring around the U.S. including Canada, Alaska (territory), Hawaii (territory), the Philippines, various locations in the Caribbean, Central America - specifically, Panama, Iceland and more.

The Service Commands are the primary organizing tool for this research and were constituted as follows:

Service Commands	
Military District of Washington	Washington, DC and environs
1st Service Command	Connecticut, Maine, Massachusetts, New Hampshire, Rhode Island, Vermont
2nd Service Command	Delaware, New Jersey, New York
3rd Service Command	Maryland, Pennsylvania, Virginia
4th Service Command	Alabama, Florida, Georgia, Mississippi, North Carolina, South Carolina, Tennessee
5th Service Command	**Indiana, Kentucky, Ohio, West Virginia**
6th Service Command	Illinois, Michigan, Wisconsin
7th Service Command	Colorado, Iowa, Kansas, Minnesota, Missouri, Nebraska, North Dakota, South Dakota, Wyoming
8th Service Command	Arkansas, Louisiana, New Mexico, Oklahoma, Texas
9th Service Command	Arizona, California, Idaho, Montana, Nevada, Oregon, Utah, Washington

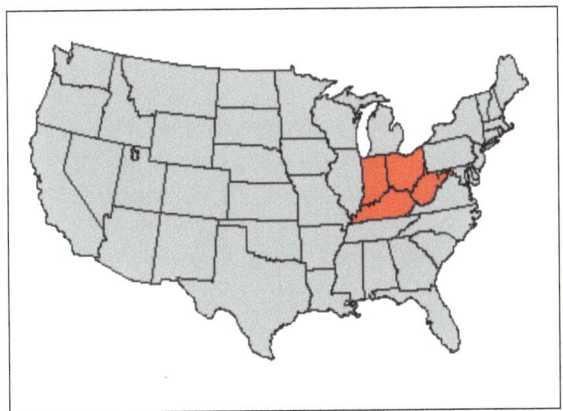

Changes to Service Commands
Before the 1943 reorganization of Service Commands:
2nd – Delaware was not listed
4th – Contained Louisiana, now in the 8th
7th – Contained Arkansas, now in the 8th; North Dakota and South Dakota not cited
8th – Arizona included; now in the 9th; Colorado included and it is now in the 7th
9th – Idaho and Nevada not cited
Furthermore, the structure was different – the nomenclature was as follows hierarchically:

Corps Areas
    Military Areas
        Military Districts

Sources of Data for the Research

Actually, there have been a great many resources upon which the researchers have drawn. These have varied based on the specific purpose to be addressed. We have attempted to identify ALL U.S. Army stations within the CONUS definition. Once someone attempts anything based on a 100% sample, the possibility of error or oversight creeps in, however, numerous sources have been consulted so as to provide at least two means of documentation for each facility so as to avoid the obvious pitfalls.

Historic (WWII) documentation and current utilization: have been developed largely through libraries (local to the site), field visits (to the site), historical societies, veterans, and very clearly - the internet. Likewise, we have attempted to take our own photography of sites visited as well as pull from the internet, post cards, and other images from books and magazines.

Special details have been garnered through the research library at the Center of Military History, Fort McNair, Wash- ington, D.C. Here, the primary need to be fulfilled was not facility descriptions or histories but unit histories. To that end we have coupled an identification of the units stationed at each facility in terms of the periodicity provided through (a) T*he Station List of the Army of the United States* and (b) T*he Directory of the Army of the United States*. These two series of documents provide the same information but organized in an opposing manner. The Station Lists, provide a listing of each unit stationed at the facility during the reporting period. The Directory provides a list- ing of units within the broader definition of U.S. Army branch and sub-branch identifying the location of the unit. These data covered Army Ground Force units, Army Air Force Units, and Army Service Force units and augment our descriptions of the installations..

In looking at Prisoner of War Camps, the same field methodology was utilized, however, documentation for this set of data was derived not from the Office of the Adjutant General of the United States, but rather from the Office of the Provost Marshal of the United States. [1]

Organization of This Volume

The principal organizing criterion for identifying the forts, camps, bases, and other stations of the U.S. Army from the World War II era is the division of the continental U.S. into the 10 service commands. There is a volume in this series for each such Service Command. Within each volume, the data are arranged according to the alphabetized list of states that comprise that service command, Within each state definition are the write-ups for the (a) Army Ground Forces, (b) the Army Air Forces, (c) the Army Service Forces, and (d) the Prisoner of War Camps. Where applicable there is attendant discussion of facilities identified as being under the direct command of the War Department and a variation on the POW camp descriptions includes that for the internment camps - largely utilized for Japanese-Ameri- cans but also used for seamen and others from Axis nations who were in the U.S. at the time of the outbreak of WWII.

Of note as well. . . there were approximately 1.6 million service men (and women) in each of the primary service branches during the war. For the most part, those in the ASF (Army Service Forces) were stationed in the U.S. Within the definition of ASF facilities, however, are numerous types of facilities only some of which were populated with sol- diers of the ASF. These included such facilities as the various bases, general hospitals, and more. In addition to ASF military at the various bases and stations, there were also non-military facilities such as ordnance works and plants, depots, etc. The database also includes those communities which had a military presence. These could have been as simple as induction centers but included a wide range of facilities identified with the community.

Footnotes: (1) These data were provided by Kathy Kirkpatrick, Salt Lake City, who conducts research in the field of genealogy with a strong orientation toward histories of Italian prisoners of war. She very graciously assisted our efforts in this regard providing the Provost Marshal General's digital records.

Station List & Directory of the Army of the United States

Both the Station List of the Army of the United States and the Directory of the Army of the United States present essentially the same information simply in juxtaposition. The primary organizing element of the Station List is the name of the post or base, or other facility – all identified as "stations." In the case of the Directory List, the primary organizing elements are the service branches, i.e., Military Police Corps, Infantry, Field Artillery, Coast Artillery, Medical Department, Transportation Corps, Quartermaster Corps, Ordnance Department, and the other such elements.

Even though the Directory has a more complete chronology it is infinitely more difficult with which to work, therefore, the data are shown by Station List orientation and augmented with data from the Directory where possible.

The War Department definition of the "war period" ran from 1 January 1939 through 31 December 1946. Information for these extended periods is sketchy and is cited where possible. The extended periods represent the build-up period prior to the advent of the war and the step-down following the war. With the surrender of Japan in August 1945, hostilities ended, however, the data for these lists continued through October of that year.

What these two types of data illustrate is an identification by unit, albeit command, division, brigade, regiment, battalion, company, detachment, or unspecified unit at which base they were stationed during the reporting period. These data reflect only those units for those time periods which were stationed within CONUS – the Continental United States. For purposes of the war, quite obviously, a great many of the units were stationed overseas.

The units reflected in these compilations are from the Army Ground Forces, the Army Air Forces, [2] and the Army Service Forces. Our documentation also reflects the activity at the various POW camps spread around the U.S.

Data for the Station List and the Directory were compiled by the Office of the Adjutant General of the United States while data for the POW camps and the POWs themselves were compiled by the Office of the Provost Marshal General.

Station List of the Army of the United States												
1941	Jan	Feb	Mar	Apr	May	Jun	Jul	Aug	Sep	Oct	Nov	Dec
1942	Jan	Feb	Mar	Apr	May	Jun	Jul	Aug	Sep	Oct	Nov	Dec
1943	Jan	Feb	Mar	Apr	May	Jun	Jul	Aug	Sep	Oct	Nov	Dec
1944	Jan	Feb	Mar	Apr	May	Jun	Jul	Aug	Sep	Oct	Nov	Dec
1945	Jan	Feb	Mar	Apr	May	Jun	Jul	Aug	Sep	Oct	Nov	Dec

Directory of the Army of the United States												
1941	Jan	Feb	Mar	Apr	May	Jun	Jul	Aug	Sep	Oct	Nov	Dec
1942	Jan	Feb	Mar	Apr	May	Jun	Jul	Aug	Sep	Oct	Nov	Dec
1943	Jan	Feb	Mar	Apr	May	Jun	Jul	Aug	Sep	Oct	Nov	Dec
1944	Jan	Feb	Mar	Apr	May	Jun	Jul	Aug	Sep	Oct	Nov	Dec
1945	Jan	Feb	Mar	Apr	May	Jun	Jul	Aug	Sep	Oct	Nov	Dec

Footnotes: (2) Prior to 1943, the Army Air Forces were known as the Army Air Corps and at the same time the Services of Supply became the Army Service Forces. Beginning in 1947, the branch was once again redesignated to be known as the United States Air Force.

Appearances suggest a more extensive database against which to draw based upon the utilization of the Directory of the Army of the United States (the Directory). However, since this documentation is based on the physical extent of the installation, the Station List of the Army of the United States (the Station List) serves as a better representation of the various army units training or stationed at any given station. Where practical, this has been augmented with citations from the Directory, however, the ultimate details relative to unit locations is the separate database developed by the authors which carriers the same title (above).

The database is written in Microsoft Access and is simultaneously provided in Microsoft Excel. The program is menu driven and contains the following components:

- Name of Station (post, base, camp, or other - military post, community, or other)
- Location (local community / state)
- Longitude / latitude of the installation
- Name of Unit
- Month(s) / Year(s) of unit assignment to station
- Applicable TOE (showing number of personnel by unit type)
- Branch of service
- Commanding Authority under which the unit was assigned
- Redesignations (name changes / assignments of the unit
- More information is constantly being added to the huge database

Fifth Service Command

Indiana

Indiana

Army Ground Forces

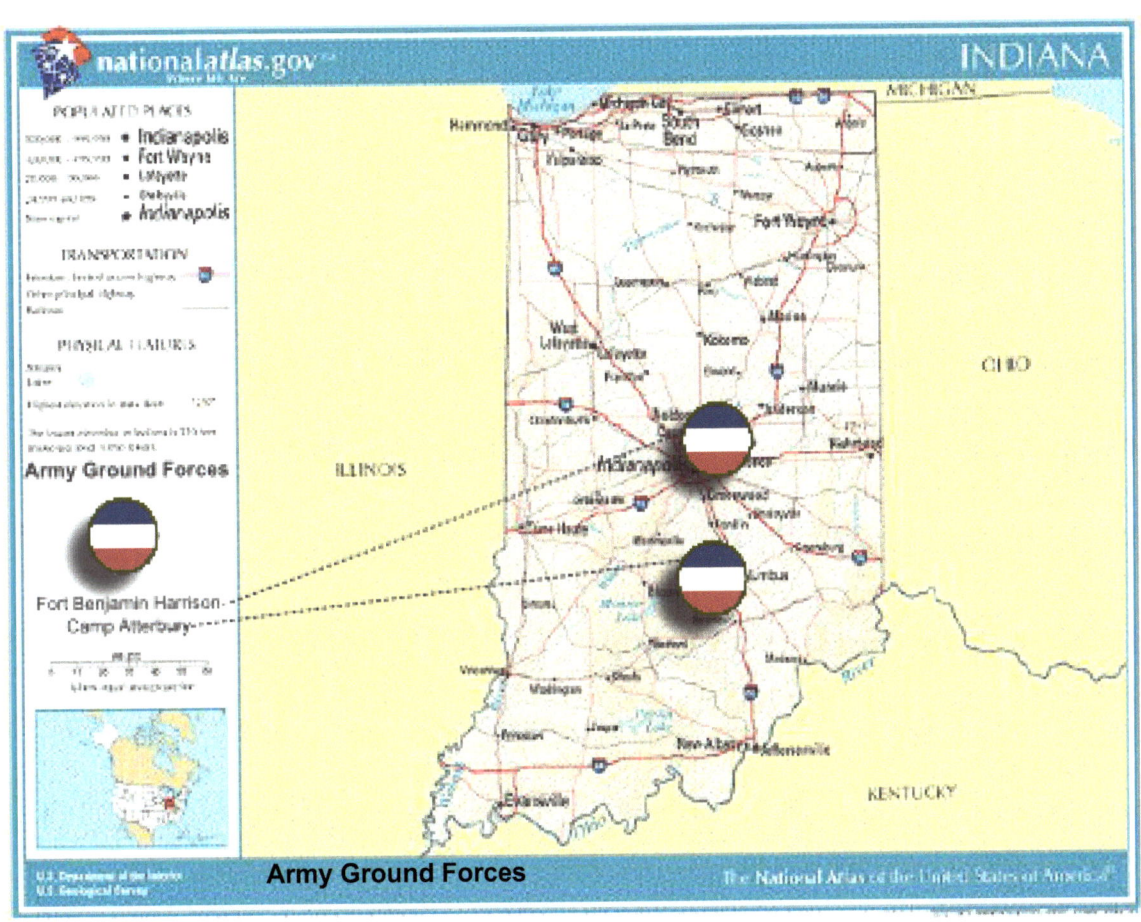

Camp Atterbury

Location	Edinburgh, Indiana *(west of I-65 on SR 252)* 39 21 10N 87 02 05W
Established	6 March 1942
Inactivated	31 December 1946 – declared inactive, but reactivated on 24 August 1950 for the Korean War
Size of Camp	40,375 Acres
Role	Established as one of the new "Triangular Division" training camps. During World War II, more than 100 units, nearly 275,000 men received their training at Camp Atterbury and others who received their initial training were sent here for advanced training. Among the units trained here were the 83rd Infantry Division, the 92nd Infantry Division, the 30th Infantry Division, the 93rd Infantry Division, and the 106th Infantry Division.
Capacity	2,389 Officers / 26,947 Enlisted personnel

Notes The post operated a major medical facility, Wakeman Hospital, for injured soldiers returning from combat as well as a POW facility for 3,000 German and Italian prisoners (another source put the number of prisoners at 15,000,.but that is unlikely - more likely it reflects the turn-overs and pass-throughs).

During World War II, more than 100 units, nearly 275,000 men received their basic training here while others were brought in for advanced training. Among the units trained here were the 83rd Infantry Division, the 92nd Infantry Division (Colored) [3], the 30th Infantry Division, the 93rd Infantry Division (Colored), and the 106th Infantry Division.

[Note:]
The term "Colored" was a term used by the U.S. Army during WWII era and is technically a part of the title of the unit (this is not in the author's lexicon).

The gymnasium at Camp Atterbury circa WWII.

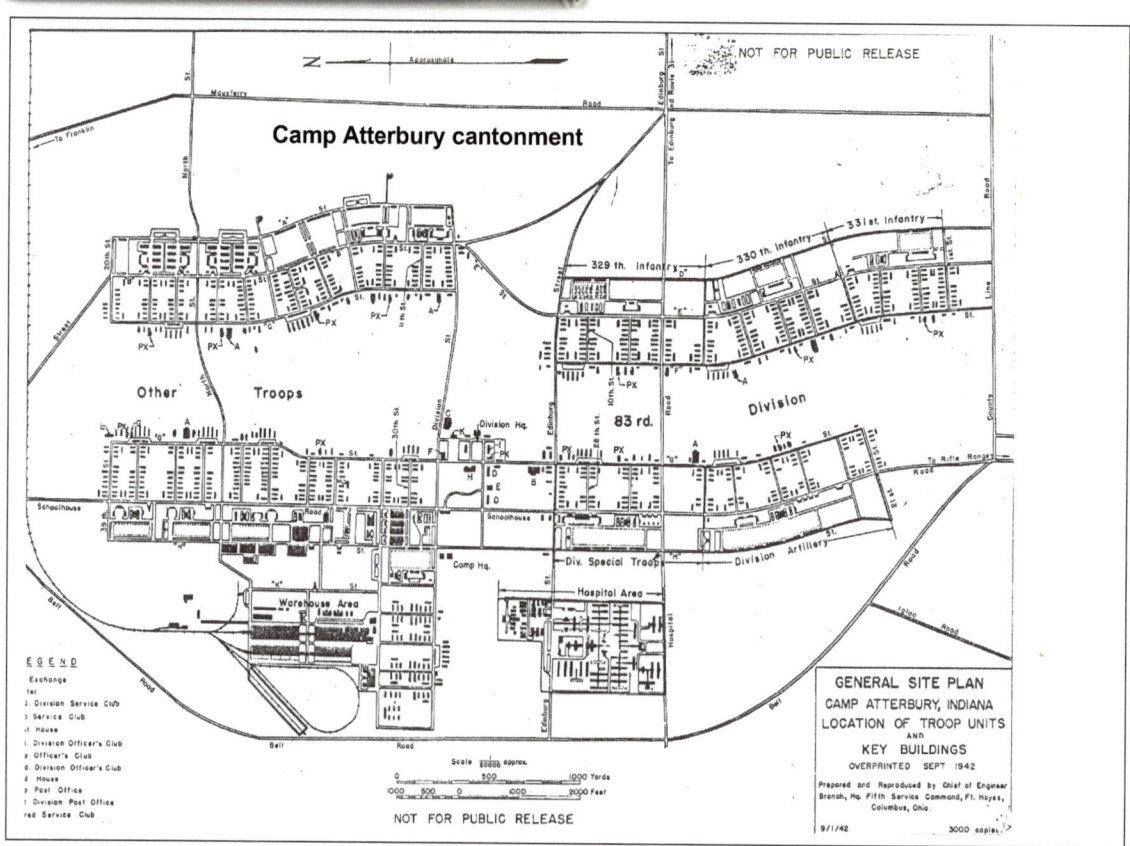

The camp included the following facilities: [3]

Enlisted Barracks (EM) – 46
WAC Barracks – 23
Officer Cottages – 6
Chapels – 12
Service Clubs – 8

BOQs – 40
POW Barracks – 61
Mess Halls – 193
PX – 16
Officer Clubs - 3

Footnote 3 – https://www.in.gov/library/about/general-information/exhibits/world-war-ii-online-exhibit/camp-atterbury--photograph-collection/
Fort Benjamin Harrison

Key Events	
2 June 1942	First General Order is published marking official activation
15 August 1942	83rd Infantry Division reactivated at Camp Atterbury
1 September 1942	8th Detachment, Special Troops, Second Army activates at post
10 September 1942	365th Combat Team (365th Infantry Regiment), 597th Field Artillery Battalin, 92nd Infantry Division is activated (full division)
20 April 1943	First Italian POWs arrive
10 November 1943	30th Infantry Division arrives from Tennessee Maneuver Area; departs 26 January 1944
27 March 1944	106th Infantry Division arrives; departs on 13 October 1944
5 April 1944	U.S. Army Station Hospital is redesignated as Wakeman General Hospital
15 October 1944	Camp Atterbury Separation Center is Activated
31 December 1946	Wakeman General Hospital is activated
31 July 1946	Separation Center inactivated
1 August 1950	Camp reactivated for training purposes for Korean War

Other elements of Camp Atterbury include:
- Atterbury Army Air Field
- Camp Atterbury Base POW Camp
- Wakeman General Hospital

Fort Benjamin Harrison

Location Indianapolis, Indiana
39 51 17N 86 00 51W
Activated 3 March 1903
Inactivated June 1947 – placed on inactive role
Size of Camp 2,415 Acres
Capacity 632 Officers / 10,719 Enlisted men; troop capacity - 14,368
Role Fort Benjamin Harrison was an infantry post during World War II housing the 10th, 11th, 20th, 23rd, 40th, 45th, and 46th Infantry Regiments. The post was established as a Reception Center for 200,000 servicemen (more or less) who were processed through the Reception Center and sent to other posts for basic training. The post was the home of the Army Finance Replacement Training Center as well as the Army Chaplain School; it

provided a cook and baker school, military police disciplinary barracks, and a POW camp for German and Italian prisoner

Postcard picture of Fort Benjamin Harrison Reception Center - WWII

Fort Benjamin Harrison post HQ

Pillar at gat to Fort Benjamin Harrison

Army Air Forces

(The following army air fields are not listed in any specific order)

Notes for Understanding Army Air Field synopses:

There were several types of army air fields, some different due to operational functionality, some being defined by ownership - frankly, finding a common listing of air fields from the WWII era is difficult.

Army Air Fields - built by the U.S. Army
Air Fields - leased by the U.S. Army
Primary Fields
Sub-Base (fields)
Auxiliary Fields
Landing Strips
Non-air field AAF Facilities
Additionally, a support field, or auxiliary field, could during one month be auxiliary to one field but be assigned to another primary field the next.

Atterbury Army Air Field

Location	Columbus, Indiana
	39 15 42N 85 53 43W
Established	February 1943 *(construction began 13 August 1942)*
Inactivated	Deactivated in 1946 until reactivated in 1949 then closed
Units	304th Army Air Force Base Unit *(host unit)*
	Initially, the field was under the control of the First Air Force, then Troop Carrier Command, transferred to the Third Air Force and then back to Troop Carrier Command.
Notes	The field was renamed Bakalar Air Force Base in the 1950s. Originally, the field was established as a sub-base to Godman Field at Fort Knox for the Third Air Force; the Tuskegee Airmen trained here (in part) in 1944.
	The field was also known as Columbus Municipal Airport [6] and as the Bartholomew County Airfield/ [4]
Capacity	1,030 troops
Auxiliaries	None
Footnotes	(4) Discussion with docents in attendance at Bakalar Air Force Base Museum.

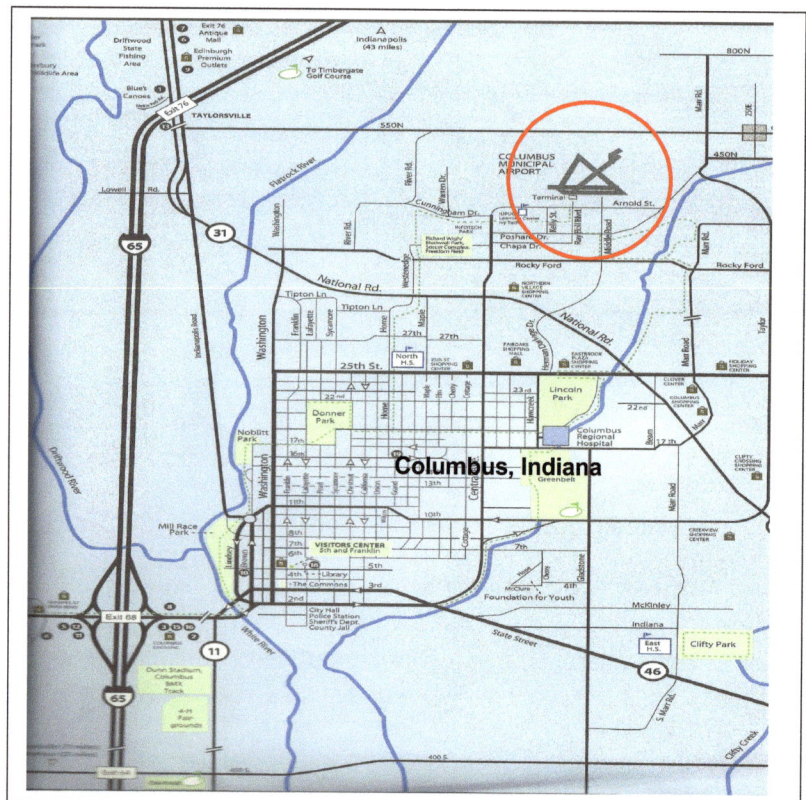

Dedication plaque outside the Chapel at the air base.

*Historic reference plaque at site
Of the army air field*

*Atterbury
Army
Air
Field
WWII*

Baer Field

Location	Fort Wayne, Indiana
	40 58 57N 85 11 16W
Established	1941 - construction begun; the Army took possession on 31 October 1941
Inactivated	31 December 1945 - placed on inactive status; Baer Field remained as a Separation Center until 10 March 1946. The base then began the process of closure: 10 January 1946 - the mess hall closed; 6 February - the finance office closed and on 29 May of that year the Post Exchange Closed. The field was declared surplus on 6 February 1947.
Role	The field was assigned to the First Air Force, headquartered at Mitchel Field at Hempstead, Long Island. The major responsibility of the First Air Force was the organization and training of bomber, fighter and other units and crews for assignment overseas.

Baer Field was subsequently assigned to the Third Air Force in March 1942. While being used by the I Concentration Command, Baer Field staged and processed medium bombardment groups. The first tactical units started to arrive for processing in September of 1942. Later, the field was assigned to the I Troop Carrier Command. Then its responsibility was the processing of tactical units of the Troop Carrier Command. The last role for Baer Field followed was as an assembly station for redeployment of personnel from the U.S. and Europe to the Pacific Theater (this role was short-lived).

Baer Field became the Fort Wayne International Airport (and was so named in 1991) although a few of the remnants can still be found from the WWII era: hangars 39 and 40 still exist in various degrees of dereliction. Some streets and building foundations can still be identified as well. During the initial phase of the field's construction, base headquarters was located on the second floor of the National Guard Armory on North Clinton Street. Everything was quite "spartan." Office furniture consisted of a homemade desk with some typewriters and one bookcase. The filing cabinets were the window sills with rocks used to hold down the papers. Later, half of a ping pong table was added along with several banquet tables from the Knights of Columbus meeting hall. Office supplies were donated by local office supply dealers.

Size of Facility 700 acres at the onset of World War II; 1,000 acres toward the end of the war; there were 3 runways each about 6,300 feet in length. The field included a chapel, barracks, fire station, hospital, administration building, warehouses, mess halls, class rooms, and repair shops - in addition to the hangars.

Capacity 3,930 personnel

Auxiliaries None

Aerial of Baer Field from 1943

looking west at 13th & "D" streets

Bendix Army Air Field (South Bend Municipal Airport)

Location	South Bend, Indiana *(3.6 miles northwest of downtown)*
	41°42'30"N 086°19'02"W
Established	Opened in 1933 and a municipal airport
Inactivated	Not known
Role	Bendix Field operated under the Air Transport Command.
Notes	The field now functions as the Michiana Regional Airport
Size of Facility	Not known
Auxiliary Field	None

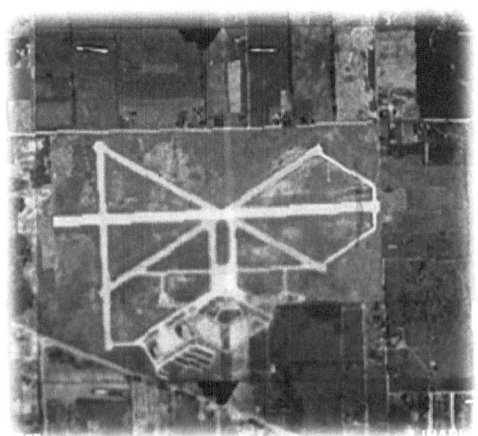 

Bendix Army Air Field (WWII) and (recent) South Bend Municipal Airport

Evansville Municipal Airport

Location	Evansville, Indiana *(3.6 miles northwest of downtown)*
	38°02'18"N 087°31'51"W
Established	Opened in 1926 and a municipal airport
	Leased by the War Department and Republic Aviation beginning in May 1942
Inactivated	Not known
Role	Evansville Municipal Airport was used for ferrying aircraft detachments during the war. In September, 1942, Republic Aviation completed the first Thunderbolt (P-47) aircraft. Some 6,075 Thunderbolt aircraft were manufactured in Evansville by Republic and delivered throughout the country and to overseas battlefronts.
Notes	Because of imminent Army takeover in 1943, the Airport Board purchased 148 acres of land on Slaughter Avenue near Burkhardt Road for an auxiliary airport. The cost of construction was $63,000 and by June all private flying was diverted to this new airport leaving only military and commercial service conducted here. Organizations that operated there included: Evansville Flight Service, Culver Flying Service, and Midwest Air Transport. The field now functions as the Evansville Regional Airport
Size of Facility	Not known

Auxiliary Field None

1940 image of Evansville Municipal Airport

Freeman Field (Seymour Advanced Flying School)

Location	Seymour, Indiana *(2.5 miles southwest of downtown)*
	38°55'29"N 085°54'30"W
Established	December 1942
Inactivated	Deactivated 1946; 1947 - deeded to the City of Seymour
Role	Freeman Field operated under the Air Technical Service Command, Southeast Training Center.
Notes	The field was used as part of the 70,000 Pilot Training Program for advanced flight training for crews of multi-engine planes. WACS (Women's Auxiliary Corps Service) and WASP's (Women's Air Service Pilots) were used for test flights.
Units	447th Army Air Force Base Unit
Size of Facility	Not known
Capacity	4,150 troops
Auxiliaries	**Grammer Auxiliary Field**
	Grammer, Indiana
	Grammer was located 18 miles north-northeast of Freeman Field.

Milport Auxiliary Field
Milport, Indiana
For a period, Milport was an auxiliary to Godman Field at Fort Knox. The field was located 15 miles to the southwest of Freeman Field.

St. Anne Auxiliary Field
North Vernon, Indiana
For a period, St. Anne Field was an auxiliary of Godman Field at Fort Knox. The field was located 18 miles to the northeast of Freeman Field.

Walesboro Auxiliary Field
Walesboro, Indiana'
For a period, Walesboro was also operating under Godman Field at Fort Knox. The field was located 15 miles to the north of Freeman Field.

Zenas Auxiliary Field
Zenas, Indiana
Zenas was located 27 miles northeast of Freeman Field.

Freeman Field and its auxiliary fields

Weir-Cook Municipal Airport (during WWII)

Location   Indianapolis, Indiana
39 43 48N 86 16 23W

Notes   Although the airport is cited in the *Station List of the Army of the United States*, no details could be identified as to the facility itself. However, it had originally been known as Mars Hill Airport and later as Indianapolis Municipal Airport, the field was named in 1944 after a WWI pilot who died in a crash in the Pacific Theater - Harvey Weir Cook. The U.S. Army Air Corps leased the field from the city for $1 per year during the war.

Postcard view of the passenger terminal at Weir Cook Airport during WW II

Stout Field

Location   Indianapolis, Indiana   *(4.5 miles SW of downtown)*
39 44 15 N 86 13 39.41W

Established   1926 - built as a municipal airport and home for the National Guard; activated for the Army Air Force in early-1942.

Role   The field was used by the Indiana Air National Guard until 1961 but is now part of an urban development area for the City of Indianapolis.
Stout Field served as the Headquarters, I Troop Carrier Command. This included a testing of gliders for air assault operations.

In early 1942 the Army Air Force leased the nearly empty airport from the state and converted it to a training field and headquarters for the newly created I Troop Carrier Command. It operated C-47, C-53, and C-46 transports.

Units   362nd Army Air Force Base Unit
I Tactical Command   20 May 1944 - 24 November 1945
IX Tactical Command   5 September 1945 - 1 February 1946

Size of Facility   200 Acres; later upgraded to 357 Acres

The top right of this photo is the intersection of Holt Road and Minnesota avenue. The air field is to the southwest of that intersection. During the war, the field occupied 3 of the 4 quadrants of the intersection.

Site photos of the former Stout Field - the remaining hangar and control tower. Below is the entrance to the Stout Field area.

U.S. Army Installations – World War II: Fifth Service Command

Schoen Field

Location Indianapolis, Indiana   *(at Fort Benjamin at the current U.S. Army Finance Center)*
 39 51 15 N 86 00 51W)
Established 1922
Size of Facility 148 acres
Role The field was used by the Indiana Air National Guard until 1961 but is now part of an urban development area for the City of Indianapolis.
Notes In 1948 Schoen Field and Fort Benjamin Harrison were jointly renamed as Benjamin Harrison Air Force Base.

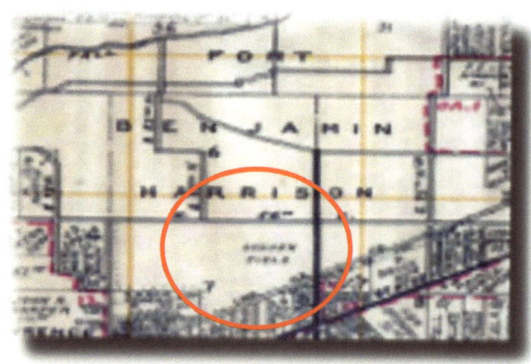

Schoen Field occupied the area that later became the U.S. Army Finance Center.

Madison Army Air Field

Location Located at the Jefferson Proving Grounds, north of Madison, Indiana
 (about 7 miles north-northwest of downtown Madison)
 38 49 43.15N 85 25 49.45W
Established 1940 - work was begun to build the Jefferson Proving Grounds (ordnance testing)
 12 July 1941 - construction begun on building the Madison AAF at Jefferson PG
Inactivated The facility reportedly closed in 1995.
Role The field was used as an operational support base for Jefferson Proving Ground which is to say, the Madison AAF tested and evaluated all types of munitions.
Notes The airfield does not show up is the common listings of AAF facilities but does record Army units to have served at the facility in the Station List of the Army of the United States.
Units 3rd Proving Ground Detachment
Size Jefferson Proving Grounds was an estimated 55,000 acres while that used for the Madison Army Air Field consisted of 637 of those acres.

The Madison Army Air Field hangar (left) at time of operations and a derelict (right).

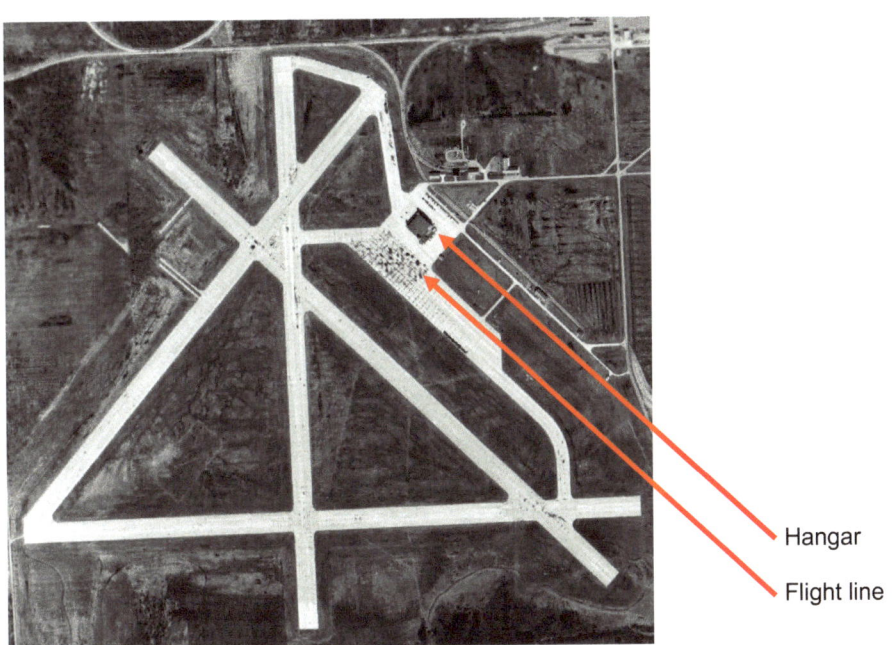

— Hangar
— Flight line

Other Air Fields   [Used by the AAF during the war but having limited information]

Location   **Bloomington Airport**
39 08 53N 86 36 55W
Bloomington, Indiana   (4.5 miles west-southwest of central Bloomington)
Now operated as Monroe County Airport

Location   **Bunker Hill Air Force Base**
40 39 53N 86 08 39W
Peru, Indiana   (12 miles north of Kokomo's downtown)
Because this is an Air Force Base, it is often misconstrued as having been an AAF field during the war, however, it was operated by the U.S. Navy until turned over to the Air Force in 1954. The field was established 1 July 1942 and contained 2,688 Acres

WWII era facilities still in Place at Bunker Hill AFB.

Location   **Paul S. Cox Airport**
Terre Haute, Indiana
39 25 21N 87 24 31W
(3 miles south of downtown Terre Haute)
In 1926 Paul S. Cox directed the formation of the Terre Haute Airport which leased 168 acres on the city's southern edge for an airfield and known as Dresser Field. The field was rededicated as Paul Cox Field on 21 June 1933 as a memorial to his leadership. The field was used by the Army and shown in the Station List as a field of use but without any army units identified.

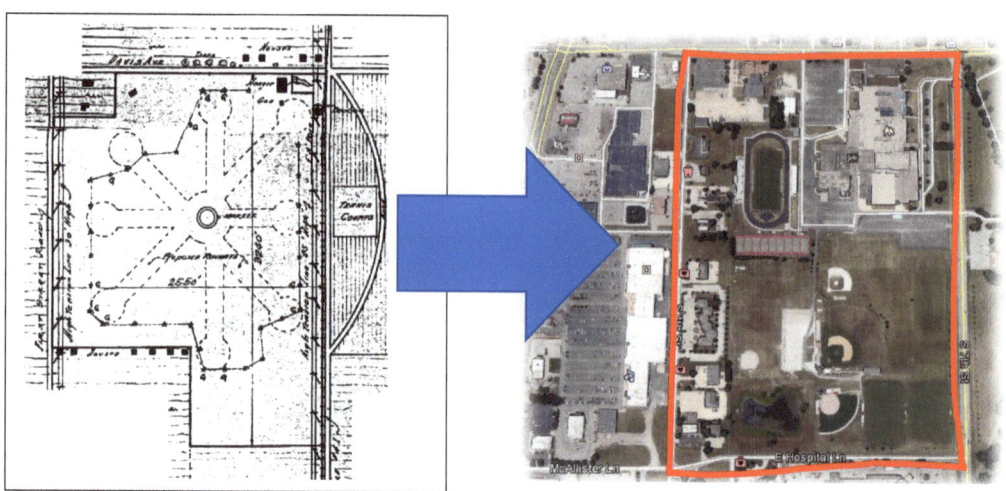

The circumscribed area shows the area consumed by Cox Field, southside of Terre Haute, Indiana.

Location   **Madison Airport**
38 49 41N 85 26 06W
Madison, Indiana   (less than 5 miles to the northwest of downtown Madison)
652 acres were purchased in January 1943 to build the airport that saw limited use during World War II.

Location   **Richmond Municipal Airport**
39 45 21N 84 50 40W
Richmond, Indiana   (5.75 miles southeast of central Richmond)
The airfield covers 702 acres.

Location   **Urschel Airport**
41 27 06N 87 00 27W
Valparaiso, Indiana   (about 3 miles southeast of central Valparaiso)

U.S. Army Installations – World War II: Fifth Service Command

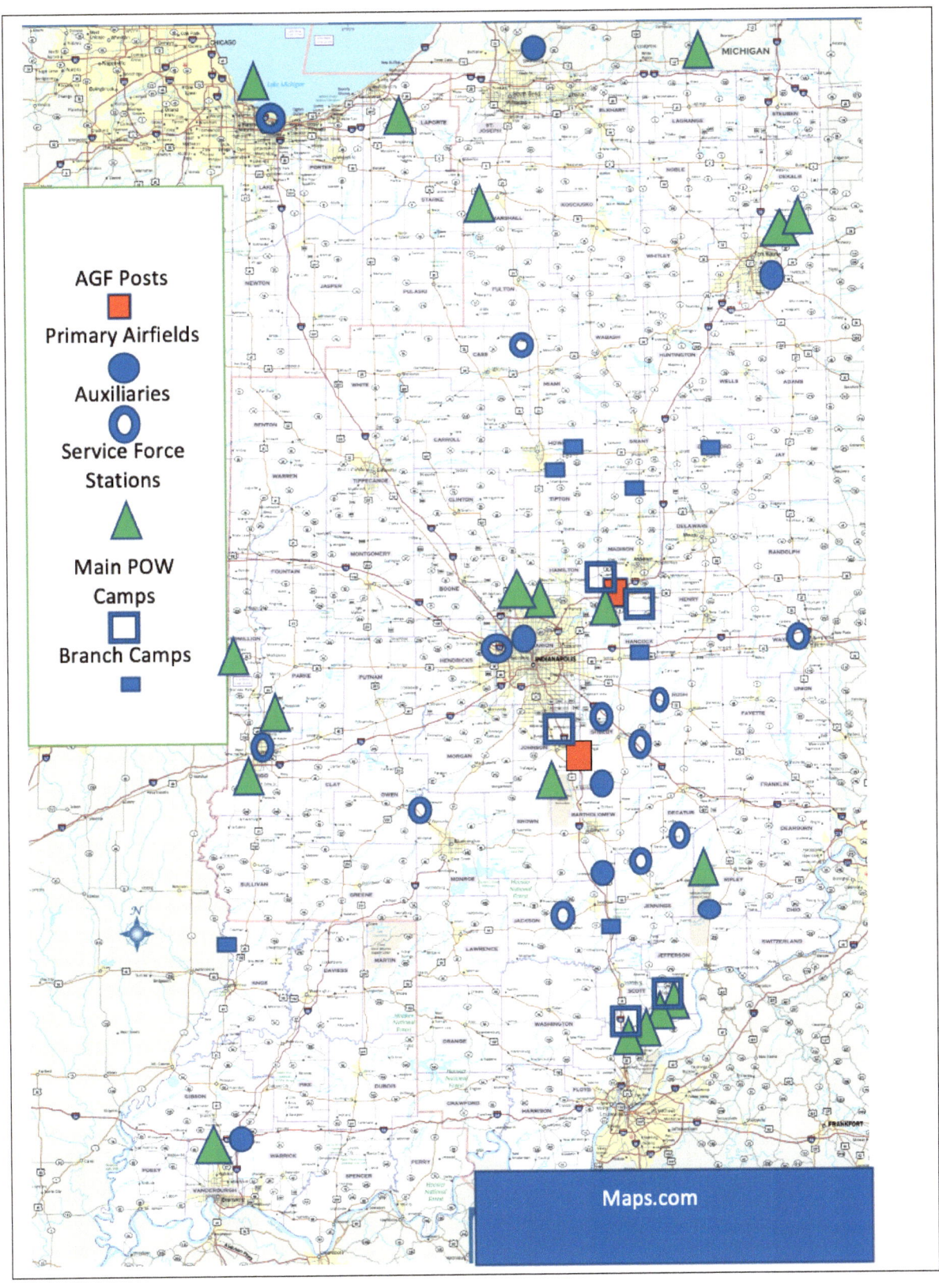

U.S. Army Installations – World War II: Fifth Service Command

Army Service Forces

Service Force units represent a wide variety of functions that are generally support provisions for tactical units in the field, e.g., medical units, quartermaster units, ordnance units, **etc**. And within these broad categories, there is a plethora of functional services provided using army posts, bases, or other stations as the focal point for the activity or they might be provided through a local community, even in the form of a commercial building.

To provide some definition to this broad range of activities and organizations, they are broken up here according to the following hierarchy:
Units providing a service from a military base
Units providing a service from a "community" base, such as a military school, a university, or other.

Camp Thomas A. Scott

Location	Fort Wayne, Indiana 341 03 16N 85 05 24.5W
Established	15 March 1942
Inactivated	1 November 1944 - converted to a prisoner of war camp
Size of Facility	79 Acres
Capacity	696 troops
Role	Camp Thomas A. Scott was established for the training of the 750th Railway Operating Engineer Battalion. The 1564th Service Command Unit later replaced the Engineer Battalion. There were several other such posts around the U.S. established for the purpose of training railway operating personnel and railway maintenance personnel for service overseas as part of the logistical support function for the troops in the field. Camp Millard in Ohio and Camp William C. Reid in Clovis, New Mexico.
Notes	Once the unit - the 750th - was trained for service and shipped overseas, the army converted the post to a prisoner of war camp which, although it was in Indiana, fell under the jurisdiction of Camp Perry in LaCarne, Ohio. Since the end of the war, the property which adjoined the railroad right-of-way has been converted into a nature habitat - almost like an oasis surrounded by railroad tracks and industrial buildings.
Units	750th Railway Operating Engineer Battalion 1564th Service Command Unit

(Left) Google Earth image of Camp Scott; (right) current use of site. Arrow points to

General Hospitals

Billings General Hospital

Location	Fort Benjamin Harrison
	39 51 31.47N 86 00 35.16W - at post flagpole
Established	1941 - constructed
Inactivated	March 1946 - closed
Size	It isn't known precisely how much of Fort Benjamin Harrison was represented by the Billings General Hospital; however, the overall post was 2,501 Acres (1,069 of which was woodland). The facility had a capacity for 2,145 troops.
Role	Billings General Hospital - Medical Department Enlisted Technicians School operated from 6 July 1942 to 17 March 1945
Notes	Billings was one of several major facilities at Fort Benjamin Harrison - included were Schoen Field, U.S. Disciplinary Barracks, POW Camp, Army Finance Center and more.

(Above) Buildings at site of Billings General Hospital (WWII); (Below) present-day use.

Wakeman General Hospital

Location	Camp Atterbury
	39 21 36.75N 86 02 59.33W - at flagpole
Established	June 1942 - as a Station Hospital; re-opened 4 April 44 as a General Hospital
Inactivated	1946
Size of Facility	75 Acres (within Camp Atterbury); 47 2-story buildings
Role	The hospital went through a considerable metamorphosis, moving from its initial designation as a Station Hospital, to a General and Convalescent Hospital and finally to the most significant designation for army hospitals of the era - Wakeman Hospital Center.
Notes	After being upgraded to a General Hospital, the facility specialized in neuro cases (brain and nerve), plastic (rebuilding body parts), and orthopedic surgery and therapeutic treatment. The hospital reached its maximum capacity of 10,000 patients in May 1945 and by the time of its closure, treated more than 85,000 patients as a general hospital.
Units	44th Evacuation Hospital
	49th Field Hospital
	118th Station Hospital
	388th Evacuation Hospital

Then and Now

Depots

Casad Ordnance Depot
New Haven, Indiana
41 04 30N 84 56 59W

Signage during wartime usage vs sign found on site in 2014.

Wartime Administration building (left); current warehousing on site. (right)

Aerial of site looking north

Location	New Haven, IN *(about 3.5 miles east of downtown)*
	41 04 21.74N 84 57 00.67W
	(at gate where Defense Logistics Agency sign found)
Established	1943
Inactivated / Closed	1945
Role	Ordnance depot; warehousing and supplies distribution

[The older photographs displayed on the preceding page were found at the Fort Wayne Public Library in McKnight, Malcolm, Casad Ordnance Depot, New Haven, Indiana, 6 November 1992]

Evansville Ordnance Depot
Evansville, Indiana
37 59 57N   87 32 50W

The plant at Evansville was operated by Chrysler which had worked out an arrangement with the Sunbeam Corporation to co-produce ammunition at one of its facilities near Evansville in the manufacture of 45 caliber cartridges for the army. The contract between the War Department and Chrysler called for the production of 7,500,000 founds of 45 caliber ammunition to be produced per week. Before the end of the first month of production, the order had been increased to 12,5000,000 units of ammunition to be manufactured.

Cartridges made at the Evansville arsenal had seven parts, passed through 48 processing operations, and had to survive 334 quality control inspections. On June 30, 1942, the first bullets produced there were test fired. From June 1942 to April 20, 1944 when the contract ended, Chrysler's Evansville arsenal produced 96 percent of the military's .45 caliber cartridges: 3,264,281,914 rounds. Rejection rate of cartridges was less then .1 percent of production.

It also produced almost a half-billion .30 caliber cartridges, hundreds of thousands of specialty rounds, reconditioned 1,662 Sherman tanks, rebuilt 4,000 Army trucks, delivered 800,000 tank grousers (track extensions for use in mud), and was preparing to make 7 million fire bombs when the war ended. [5]

Footnotes (5) http://www.defensemedianetwork.com/sgories/bullets-by-the-billions-chrysler=switches-world-war-ii-production-from-cars-tp-cartridges/

The WWII plant

Fall Creek Ordnance Depot
Indianapolis, Indiana (21st and Northwestern Avenue)
39 47 44N 86 10 06W
No details available.

Indiana Army Ammunition Plant [6]
Charlestown, Indiana
38 07 30N 85 35 52W
The INAAP, when completed, covered 10, 655 acres, and consisted of 1,700 buildings, eighty-four miles of railroad track, 190 miles of road, and thirty miles of fence. The final cost of the plant was over $133.4 million dollars. In 1989 the US Army placed the "replacement value" of the INAAP at $1.9 billion dollars.
Component Elements:

Indiana Ordnance Works Plant #1
Indiana Ordnance Works Plant 1, though government-owned, was operated by the DuPont Company. The INAAP represented a chance at a new life for the thousands of Hoosiers, Kentuckians, and other Americans who had been hit by the Great Depression. According to Julius Hock, a former employee at IOW#1, DuPont paid its unskilled laborers sixty cents per hour. Construction officially commenced on August 26, 1940, but actual building did not get underway until September 4. By December 1940 the population of Charlestown had swelled to over 13,400, with some 10,000 working at the INAAP. By May 1, 1941, employment at the plant reached its height of 27, 520 persons. Hock characterized the plant during construction as "a beehive."

Hoosier Ordnance Plant
On February 5, 1941, construction began on the Hoosier Ordnance Plant, a load, assembly, and pack (LAP) facility which was used to prepare cannon, artillery, and mortar projectiles. The plant went into operation on September 2 of that year, and construction was completed by January 31, 1942. The operation of HOP was handled by Goodyear Tire and Rubber Company.

Indiana Ordnance Works Plant #2
Work did not begin on IOW#2, the rocket propellant plant, until December 1944. Though the construction of the plant was never completed, production did take place there for about five weeks before the facility was shut down shortly after Japan's surrender in August 1945.

For a short time at IOW#2, German prisoners of war were used to supplement the labor force, chiefly as unskilled labor, such as digging ditches. The first POWs arrived in May 1945, but were quickly repatriated at the end of the war three months later.

Indianapolis Chemical Warfare Depot
Indianapolis, Indiana (2060 Northwestern Avenue)
39 43 37N 86 10 01W
No details available.

Jeffersonville Quartermaster Depot
Jeffersonville, Indiana
38 16 59N 85 44 25W
In 1871 the U.S. Army decided to build an edifice that would contain all the individual units that had spread all around Jeffersonville. Quartermaster General Montgomery C. Meigs designed the structure, which opened in 1874. Frederick Law Olmsted also helped design the facility, and much of his vision still exists

Footnotes (6) "Parent" organization for the 3 munitions plants.

with its brick structures and arched glass portals, but more of Meigs' vision won out. A 100-foot (30 m) tower was initially built as both watchtower and water tower, but was razed in 1900 to make the power plant into a two-story headquarters building.

During the Spanish-American War, 100,000 uniform shirts a month were produced, and in World War I 700,000 of the shirts were made. This gave the depot the nickname "America's largest shirt factory". In World War II the depot produced $2.2 billion in goods for the war effort. It stayed in operation for the Korean War, but by 1957 it was decided to close the facility, which happened in 1958. [7, 8]

Footnotes
(7) *This discussion is drawn from multiple sources and presented in an article under the authorship of: Vest, Rob, "Charlestown, IN and the Indiana Army Ammunition Plant: The Making of A War-Industry Boom-Town" and documented at website: http://homepages.ius.edu/ RVEST/INAAP.htm*
(8) *http://en.wikipedia.org/wiki/Jeffersonville_Quartermaster_Depot*

Terre Haute Ordnance Depot
Terre Haute, Indiana
39 30 13N 87 21 25W

Construction on the Terre Haute Ordnance Depot began on June 4, 1942. It was one of two ordnance depots activated in Vigo County, Indiana that year. The depot was completed on December 4, 1942 at a cost of $5.6 million. Terre Haute Ordnance Depot was located on 523 acres (2.12 km2) east of Fruitridge Avenue in Terre Haute, Indiana. [10] The Terre Haute Ordnance Depot was self-sustaining, storing spare parts and equipment, repairing and renovating transport vehicles, and shipping supplies to military installations. Its character remained static while the Vigo Plant underwent drastic changes. Munitions production ceased at the Vigo Plant on June 30, 1943, and control of the facility was returned to the Army Corps of Engineers. On May 8, 1944 - coinciding with a $2.5 million federal appropriate to fund a clandestine biological arsenal - the Vigo Plant was directed to convert to a facility capable of manufacturing vaccines and either 275,000 botulinum toxin bombs or one million anthrax bombs each month, ostensibly "an animal breeding laboratory." [9]

Building at the site of the former Terre Haute Ordnance Depot

Vigo Ordnance Plant
Vigo, Indiana
39 25 01N 87 27 05W

The Vigo Ordnance Plant was located on 700 acres of a more than 6,000-acre government-owned tract and cost $21 million to build.[The facility was constructed near the small community of Vigo, Indiana, about six miles south of Terre Haute. The area surrounding the plant was flat, covered with cornfields and dotted by hog farms.

The U.S. Army Ordnance Corps constructed the Vigo Plant in 1942, prior to the official start of the U.S. biological weapons program. The Vigo Ordnance Plant began producing conventional explosives and munitions on February 18, 1942. The Army decommissioned the factory less than year later, and on June 30, 1943 the plant was transferred to the U.S. Army Corps of Engineers. Portions of the Vigo Plant were then leased to the Delco Radio Corporation for the manufacture of military electronics equipment. The plant served in this capacity until May 1944.

On May 8, 1944 the Army Special Projects Division (SPD) directed the Vigo Plant to convert its facilities for full-scale biological agent production. The plant was converted for biological

Footnotes (9) *McCormick, Mike, Terre Haute: Queen City of the Wabash Google Books, 2005, p. 129*
warfare (BW) use by the H.K. Ferguson Construction Company; they added fermenter tanks, slurry heaters, laboratories and the other necessities of a biological warfare facility. The plant was to be the first U.S. anthrax factory and would be utilized filling a British order for anthrax bombs. In March 1944 the British had placed an order for 500,000 of these bombs which Winston Churchill, remarked, should only be considered a "first installment".

When it was conceived, the initial plan was for the Vigo Plant to be a production facility for anthrax and botulinum toxin. The 1944 order converting the plant to a BW facility directed that it become a factory capable of producing 275,000 boutlin bombs or one million anthrax bombs per month. The core of the Vigo Plant's BW operation was the anthrax fermenters installed during the renovations in 1944. There were 12 - 20,000 gallon fermenter tanks at Vigo, the total of 240,000 gallons which made it the largest bacterial mass-production line anywhere in the world at the time. [10]

Wabash River Ordnance Depot
Newport, Indiana
39 48 27N 87 29 41W

The Newport Chemical Depot (NECD), opened in 1941 as the **Wabash River Ordnance Works**, was one of nine Army installations in the United States that stored chemical weapons. Various warfare materials, including chemical agents, were manufactured at the site. In the late 1960's the United States stopped production and shipment of chemical agent and weapons.

The facility was located less than 3 miles south-southwest of Newport, Indiana, a short distance west of the Wabash River. In some references, the town of Dana is cited as the location for the facility.

On November 14, 1941, authorization was granted for the construction of an RDX facility two miles south of Newport, Indiana. The E.I. Du Pont de Nemours & Company, Incorporated (Wilmington, Delaware) was awarded the construction contract for the construction of a five-line RDX facility in December 1941. Construction of this RDX Manufacturing Area (RDX-MA) comprising approximately 300 acres in the north central portion the site, was completed in October 1943 at a cost of $45,717,500. In 1951, while Du Pont was manufacturing heavy water, the Liberty Powder Defense Corporation of East Alton, Illinois, rehabilitated two of the five RDX lines and related facilities at a cost of $4,361,652. Liberty Powder Defense Corporation operated the plant under contract with the U.S. Army from August 1951 until March 1957. During the period 1957 through 1960, there was no production at the site. [11]

Footnotes (10) *http://en.wikipedia.org/wiki/Vigo_Ordnance_Plant*
(11) *http://www.globalsecurity.org/wmd/facility/newport.htm*

Kingsbury Ordnance Plant
Kingsbury, Indiana
40 30 45N 86 41 02W

As it seemed more certain that the United States would become involved in World War II, the War Department began seeking locations for ammunition production. To meet the demand, Kingsbury was slated for construction in November 1940 to be one of the principal shell-loading facilities. After the plant was announced locally, it was revealed that 250 families would be displaced by the plant's new footprint; they were given 30 days to move and $113 for each sacrificed acre. Though it may sound unreasonable, this methodology was not uncommon (nor unchallenged).

On a swath of 13,000 acres, nearly 21,000 employees worked factory lines every hour of the day measuring and pouring explosives into artillery shells, bombs, land mines and grenades of all shapes and sizes.

In August 1941 Kingsbury's first shells were loaded and en route to the front lines, before the ammo facility was even complete, and just as the Germans began their siege of Leningrad. Construction completed in late February, 1942 and the plant was soon operating at full capacity, but not everyone got along.

After Japan surrendered, Kingsbury almost immediately went into standby. In 1951 shell production resumed for a short time to fulfill the needs of the Korean War, but by the end of the 1950s its services were no longer needed. The plant was permanently closed and sold off in pieces, much of it returning to its original state as farmland. [12]

Schematic Layout of Kingsbury Ordnance Plant.

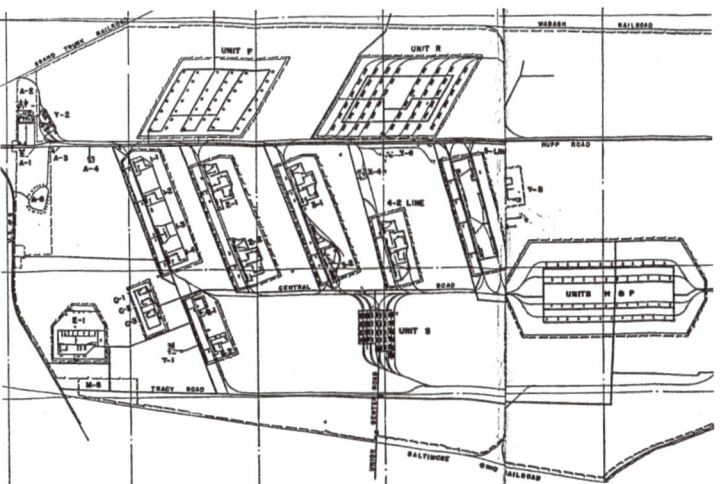

Ordnance Depots & Plants

Schools

Culver Military Academy
Culver, Indiana
41 13 17N 86 24 26W
The institution was established in 1894 "for the purpose of thoroughly preparing young men for the best colleges, scientific schools and businesses of America."

The author's preconception regarding the institution was that it trained young men (and now young women as well) toward careers in the military of the United States. However, in reviewing a listing of the notable individuals to have graduated from the Culver Military Academy, no prominent military officer was identified - no military at all. However, the listing of notables included a fascinating array of men who have been prominent leaders in their respective fields over the years. Such individuals would be listed here but the list is too long for a comprehensive inclusion of all and to list some but not all would be more of a problem.

Given this condition, it is questionable as to why the Culver Military Academy is listed among the Army Service Forces' installations in this documentation. The answer is that the *Station List of the Army of the United States* listed the school, ergo, we have included it here.

Serving at the school (see Appendix) was the:

3510th Service Command Unit: ROTC - Culver Military Academy

3 views of the campus at Culver Military Academy

Howe Military School (Academy)
Howe, Indiana
41 43 34N 85 25 31W

In that there are two military schools in northern Indiana no doubt draws comparisons between them. Such an extension of this research is beyond our intent and any observations based on our research would be of a pedestrian nature. In simplest terms, the campuses have a different feel to them - one pastoral, the other more "spartan" and of a more highly military orientation.

Like Culver, Howe is a private boarding school, now co-educational. The facility sits on a 150-acre plot very near the Michigan-Indiana state border and established in 1884 as the Howe Grammar School and evolved into The Howe School as a preparatory facility for young men seeking ordination to the priesthood of the Episcopal Church. [13] The purpose of the school was to "provide a quality education and to include the ROTC program." [14]

Footnotes (13) *John B. Howe drafted the 1851 Constitution of the State of Indiana.*
           (14) *Statement provided by Vaugh, Karl, Howe Military School*

The school became a military school in 1884, and fully co-educational in 1988, with Company A (Alpha) being the all-female company consisting of day students and those that live on campus full-time. The school has an enrollment of 98 and a faculty of 15.

As of September 2008, Howe was one of 28 military schools in the United States, down from a high of 125 such schools. Howe is designated a military institute JROTC program and has earned the designation of "Honor Unit with Distinction."

Serving at the school (see Appendix) was the:
3513th Service Command Unit: ROTC - Howe Military School

(Top) Aerial view of the campus (looking northeast) - Source: http://www.thehoweschool.org/the-campus.aspx; (Below) two views of the campus

Other Facilities

Jefferson Proving Ground
Madison, Indiana   (7 miles north-northwest of downtown Madison)
38 49 45N 85 24 49W
Jefferson Proving Ground, is a 55,265 acre facility established in December 1940; JPG's primary mission was to perform production and post-production tests of conventional ammunition components and other

ordnance items and conduct tests of propellant ammunition/weapons systems and components for the U.S. Army. It fired its first round 5 months after opening, and operated until 1995.

The majority of the land adjacent to JPG is agricultural and rural residential. JPG was constructed in 1940-41, just prior to U.S. involvement in World War II. In 1940 the Chief of Ordnance, U.S. Army Services Forces, determined the need for a large proving ground to simultaneously conduct research while performing development and production acceptance tests. JPG was constructed because existing U.S. Army proving grounds were determined to be inadequate to support the perceived World War II effort.

Test operations at the installation conducted during the World War II period consisted of production acceptance tests of ammunition and weapon systems, and their components. Test munitions included high-explosive projectiles, propellants, cartridges, primers, fuses, boosters, bombs, and grenades. Over 7 million rounds have been fired since the initial test in 1941 through September 1945.

After World War II, testing activities were sharply reduced and in March 1946, U.S. Army officials discounted JPG's status as an independent command and included it as a sub-post under the control of the Indiana Arsenal, placing most of JPG on standby status. Special production engineering tests as well as research and development (R&D) tests were conducted during the 1951 to 1955 time period. In July 1958,

JPG was again placed on a standby status, but with ammunition test capabilities held at a high level of readiness. JPG remained on such status until 1961. [15]
Chronology of JPG operations: [16]

Footnotes   (15)   *http://www.jpgbrac.com/history/history.htm*
             (16)   *http://www.in.gov/idem/4231.htm*

- 1941 - Began operations.
- 1944 - 100,000th round fired.
- 1953 - Peak staffing (1,774 employees) and 24 hour continuous testing during the Korean Conflict.
- 1960's - Peak of testing activity during the Vietnam War.
- 1989 - Congress identified JPG for closure (421 employees).
- 1991 - Peak of testing activity during the Persian Gulf War.
- 1993 - Jefferson Proving Ground Redevelopment Board (JPG RDB) formed.
- Photo of last round fired on Sept. 30, 1994 (select image for larger version)1994 - Restoration Ad visory Board (RAB) formed. Munitions testing ceased.
- 1995 - Local Redevelopment Authority (LRA) formed. JPG closed.
- 1996 - Property disposal process began: environmental site assessments and cleanup, unexplod ed ordnance (UXO) removal actions, and property lease/transfer.
- 1997 - Memorandum of Agreement established between The Army Test and Evaluation Command (TECOM) and the U.S. Fish and Wildlife Service for "ecosystem-based" management of 51,000 acres in the northern part of the base.
- 1998 - Memoranda of Understanding established between the U.S. Department of Army and the Indiana Air National Guard. In exchange for continued use of the Jefferson Range air-to-ground training area, the Air National Guard was to provide assistance to the Army in the operations and maintenance of the 51,000-acre impact area.
- 2000 - Big Oaks National Wildlife Refuge created as an overlay refuge.

The facility is divided into a northern impact area and a southern containment area, separated by a firing line consisting of 268 gun positions for the testing of ordnance. This line runs east-west across the width of the facility. The northern area consists of 51,000 acres of undeveloped and heavily wooded land.

Numerous, discrete areas in this part of the facility have been cleared and are targeted during certain munition tests. The southern containment area houses the support facilities used for administration, ammunition assembly and testing, vehicles and weapons maintenance, and residential housing. Most of these buildings are situated along a 1-mile-wide strip just south of the Firing Line Road (also known as Main Front Road).

An abandoned airport with five runways and a hangar building are located in the southwest corner of the facility. Jefferson Proving Ground contained 379 buildings, 182 miles of roads, and 48 miles of boundary fence line.

The installation is approximately eighteen (18) miles long (north-south) and five (5) miles wide (east-west)
Source: Google Earth

Gary Armor Plate Plant
Gary, Indiana   (2nd & MIssissippi Streets)
41 35 30N 87 19 18W (approximate)

The only specific item found on the wartime facilty known as Gary Armor Plate Plant was the attached article which points to the facility being operated by Carnegie-Illinois Steel Corporation in the production of tanks.

Gary Armor Plate Plant (IV) shows up in the Station List of the Army of the United States but without a specific unit assigned.

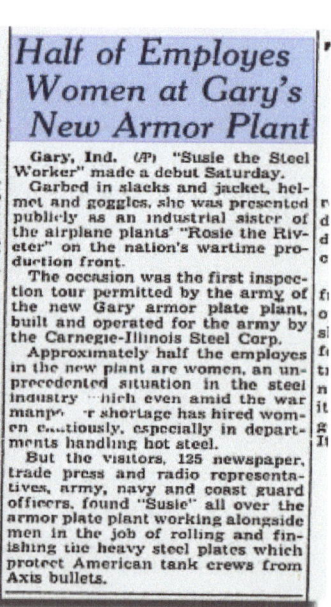

Half of Employes Women at Gary's New Armor Plant

Gary, Ind. (AP) "Susie the Steel Worker" made a debut Saturday.

Garbed in slacks and jacket, helmet and goggles, she was presented publicly as an industrial sister of the airplane plants' "Rosie the Riveter" on the nation's wartime production front.

The occasion was the first inspection tour permitted by the army of the new Gary armor plate plant, built and operated for the army by the Carnegie-Illinois Steel Corp.

Approximately half the employes in the new plant are women, an unprecedented situation in the steel industry which even amid the war manpower shortage has hired women cautiously, especially in departments handling hot steel.

But the visitors, 125 newspaper, trade press and radio representatives, army, navy and coast guard officers, found "Susie" all over the armor plate plant working alongside men in the job of rolling and finishing the heavy steel plates which protect American tank crews from Axis bullets.

Army Service Force Locations

Location	Branch	Type Facility
Anderson	Districts	Armed Forces Induction District
Bloomington	Army Service Forces	Service Command Unit: ASTU - Indiana University
	Army Service Forces	Service Command Unit: ROTC
	ASF Schools	ROTC Schools
	Schools	Civilian Schools
Charlestown	Depots	Plants and Works
Clinton	Depots	Plants and Works
Evansville	Depots	Plants and Works
	Districts	Armed Forces Induction Districts
Fort Wayne	District	Quartermaster Corps Procurement District
Gary	Army Service Forces	Service Command Unit: ROTC
	ASF Schools	ROTC Schools - Gary High Schools
	Districts	Armed Forces Induction District
Indianapolis	Army Service Forces	Service Command Unit: Finance Office, U.S. Army
		Service Command Unit: Corps of Military Police, District Offices
		Service Command Unit: District #3, Fifth Service Command
		Service Command Unit: Indianapolis Officer Procurement District
		Service Command Unit: Indianapolis Recruiting District
		Service Command Unit: Armred Forces Induction Station
		Service Command Unit: ROTC, Indianapolis High Schools
		Service Command Unit: Service Group, 836th AAF Spec. Depot
	ASF Schools	ROTC - Schools
	Depots	Plants and Works
	District	Officer Procurement District
	Districts	Armred Forces Induction Districts
	Finance Department	Finance Offices, U.S. Army
Jeffersonville	Army Service Forces	Service Command Unit: Finance Office, U.S. Army
		Service Command Unit: Jeffersonville Quartermaster Depot
	Depots	Quartermaster Depots
	District	Quartermaster Corps Procurement Districts
	Finance Department	Finance Offices, U.S. Army
	Pools	Quartermaster Department Officer Replacement Pool
Kokomo	Districts	Armed Forces Induction Districts
La Porte	Depots	Plants and Works
Lafayette	Army Service Forces	Service Command Unit: ROTC - Purdue University
	ASF Schools	ROTC Schools
	Districts	Armed Forces Induction Districts
	Schools	Civilian Schools

Army Service Force Locations

Madison	Army Service Forces	Proving Grounds
Muncie	Army Service Forces	Service Command Unit: ASTU - Ball State Teachers College
	Schools	Civilian Schools
New Haven	Army Service Forces	Service Command Unit: Casad Ordnance Depot
	Depots	Ordnance Depots
Richmond	Districts	Armed Forces Induction Districts
Seymour	Army Service Forces	Service Command Unit: Service Group, Freeman Field
South Bend	Districts	Armed Forces Induction Districts
Terre Haute	Army Service Forces	Service Command Unit: ASTU *
		Service Command Unit: ROTC, Rose Polytechnic Institute
		Service Command Unit: Terre Haute Ordnance Depot
	ASF Schools	Rose Polytechnic Institute
	Depots	Ordnance Depots
		Plants and Works
	Districts	Armed Forces Induction Districts
	Schools	Civilian Schools
West Lafayette	Army Service Forces	Service Command Unit: ASTU - Purdue University

* The Army leased office space in buildings throughout many cities around the country where they provided many of the administrative functions of their "business." These were the non-base or non-post activities where they performed like a contractor or a lessee. For example, at various universities they provided the ASTU (Army Specialized Training Unit) function and at even more they operated the ROTC (Reserve Officer Training Corps) programs. They did not maintain a "base" at the university, but staff was assigned to the university for the performance of the program requirements.

Prisoner of War Camps

There are several classes of prisoner of war camps. The camps can also be distinguished by their location, whether on a military facility or "at large" in a community.

Functional Classes
Main Camp
Branch Camp
Work Camp
Penal / Medical Institution (P/MI)

Locational Classes
Military Post / Base (or other i.e., General Hospital)
Main Camps / Off Military Bases
Branch Camps / Off Military Bases

To make tracking the data even more difficult, a camp that is identified as a Main Camp at one point in time could easily be a branch camp to some other Main Camp at another time. Mostly, but not always, a branch camp is not on a military facility, but that "rule" doesn't always apply. It's a veritable mix and match set of definitions to determine how to classify any given camp at any given time.

The data for the camps in terms of locations and prisoner counts was developed by the Office of the Provost Marshal General. Their records have been put onto microfilm and are thusly available. However, numerous pages of the data are unreadable due to one or more of the following conditions:
- pages have yellowed, type becoming less distinct, and pages flaking away
- original material was typed on an old mechanical typewriter with multiple carbons and it is uncertain which copy had been used with which to make the "official" version that was microfilmed
- the type, itself, is sometimes virtually indistinguishable or smudged
- some data have been omitted

Camp Atterbury Base POW Camp

Camp Atterbury, Indiana
39 22 31N 86 04 08W

Type Facility	Camp Atterbury Base POW Camp
Period	July 1943 through May 1944 - Italian POWs
	June 1944 through June 1946 - German POWs
Location	Camp Atterbury (Edinburgh, Indiana)
Size / Capacity	Capacity varied: 3,000 - 4,500 - 3,000
Num. Prisoners-Avg / Max	4,931 German POW average over 510 days / 7,939 peak in Nov. 1945
	2,380 Italian POW average over 300 days / 3,002 peak in October 1943
Nationality	Italian (first); replaced by Germans
MP / Service Units	429th Military Police Escort Guard Company (ASF)
	577th Military Police Escort Guard Company (ASF)
	1537th Service Command Unit: Prisoner of War Camp
Branches	Austin Branch Camp, IN
	Operated 577 days between September 1944 and April 1945; it held German prisoners an average of 685 over the 577 days with a peak of 1,642 in October 1945.

Eaton Branch Camp, IN
> Operated for 61 days holding a total of 630 German prisoners between September and November 1945

Morristown Branch Camp, IN
> This branch held a total of 593 German prisoners for 396 days between October 1944 and November 1945

Vincennes Branch Camp, IN
> At this branch 300 German POWs were held over a period of 365 days between October 1944 and October 1945

Windfall Branch Camp, IN
> Windfall, a tent camp, held 1,790 German POWs for a period of 365 days between 1944 and October 1945

Crile General Hospital Branch Camp, Cleveland, OH
> The POW facility at Crile G.H. held 225 German POWs for approximately 31 days in March-April 1946

Fletcher General Hospital Branch Camp, Cambridge, OH
> The POW facility at Fletcher G.H., likewise, held 172 German POWs for a 31 day period in March-April 1946

Marion Engineer Depot Branch Camp, Marion, OH
> The Engineer Depot at Marion, Ohio, held 210 German POWs during a one-month period in March 1946

Wright Field Branch Camp, Dayton, OH
> The facility at Wright Field was the detention center for 783 German POWs between March and April 1946

Work Camps
- Alexandria Work Camp, IN
- Marysville Work Camp, IN
- Tipton Work Camp, IN

Other
Other camps in Indiana fell under the jurisdiction of base camps in other states

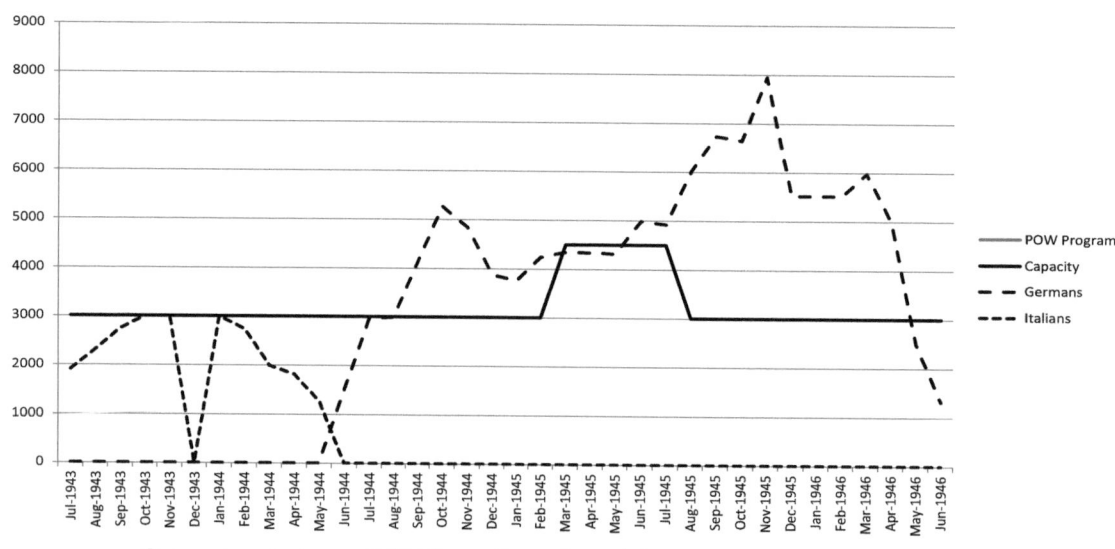

Comparison between POW Camp Capacity and the numbers of German & Italian POWs

This placque is posted outside the on-post chapel built by the Italian POWs. It has been nicely preserved and is one of the testimonials to the presence of the Italians at Camp Atterubry.

(Top) view of the eastern edge of the POW camp site; (Center) a small chapel built on site by the Italian POW - continues to be maintained quite nicely; (Below) center point of the southern perimeter of the POW camp boundry.

Italian POWs crowding the chapel they had built and shown above.

Fort Benjamin Harrison Base POW Camp

Fort Benjamin Harrison, Indiana

	39 51 34N 86 01 13W
Type Facility	Base POW Camp
Period	February to May 1944 - Italian POWs (90 days)
	June 1944 through February 1945 - German POWs (245 days)
Location	Fort Benjamin Harrison (Indianapolis, Indiana)
Size / Capacity	Capacity varied: 500
Num. Prisoners-Avg / Max	499 German POW average over 270 days / 975 peak in September 1944
	248 Italian POWs average over 120 days / 250 peak in February 1944
Nationality	Italian (first); replaced by Germans
MP / Service Units	2570th Service Command Unit: Prisoner of War Camp
Branches	None

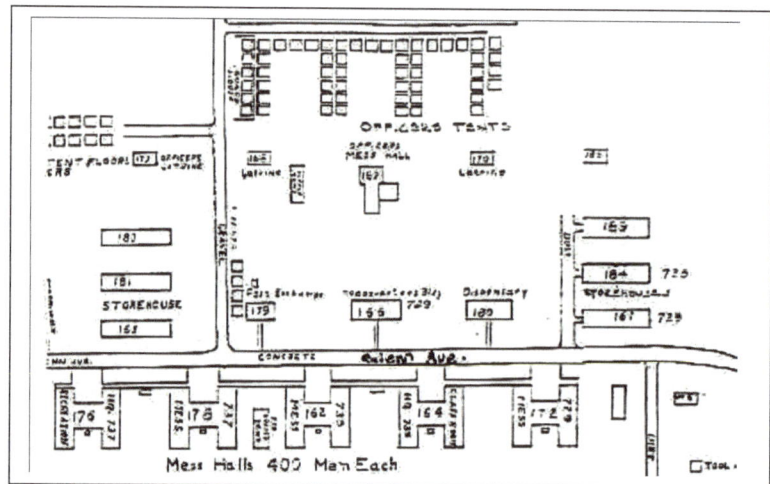

The POW camp sat astride Glenn Road (the street from the bottom of the picture with a linear parking lot in the middle. The barracks were to the left (as viewed here), and the dispensary, canteen and other facilities were to the right.

The POW camp was located in the Camp Glenn area. The interpretive center building (current use) was used as a barracks. The compound was encircled by a double fence topped with barbed wire.

In January of 1944, 250 Italians POWs from Camp atterbury moved to Fort Harrison. The Camp Edwin F. Glenn site, used for the Citizens' Military Training Camp in the 1930's was fenced and converted into a copound for POWs. Originally, two of the three mess halls were turned into living quarers for the prisoners, and the third became their mess hall. After September of 1943, the Italians had switched sides and became our allies, so by the time they came to Fort Harrison, Italian soldiers were already reclassified as "co-belligerents." This made their internment somewhat paradoxical, since they were enemies when captured, but remained prisoners even after these circumstances changed.

The Italians performed maintenance duties on the post, clean-up activities, and made improvements to the prison compound. They were at Fort Harrison only four months when they were transferred to Fort Hayes, Ohio.

In May 1944, 300 German prisoners from Rommel's Afrika Korps moved into the camp. The Germans were extremely proud and patriotic. They arrived still wearing their worn and faded military uniforms. They left a lasting impression with people who witnessed them "goose-stepping" in formation to their work site.

One story about their tenture at the post had the Germans assigned to installing a new roof on the canteen. They carefully divided two shades of shingles and placed them on the roof to form a large swastika. The staff did not discover their handiwork for several days, after which time the job had to be re-done.

German prisoners were responsible for much of the work on the new Officers Club, now the Garrison Restaurant, They also maintained and repaired other post buildings, performed road work and assisted the quartermaster with laundry, equipment and motor pool duties. Some even waited tables at the Officers Club.

Fort Harrison later became a site for an Army Disciplinary Barracks, a prison compound for American servicemen convicted of offenses by the military court system. Because Army regulations did not allow POW camps in the same military installation that housed disciplinary barracks, Fort Harrison's POW camp was closed in February 1945 and the German prisoners moved to the facility at Fort Knox, Kentucky. [17]

Footnotes   *(16)   Source: The article (as edited) was provided by the Indiana Department of Natural Resources who manages the grounds that once were where the POW camp had been established. Coincidentally, the author was told by the staff at the facility that taking pictures of these state grounds was forbidden, even passage onto the facility was forbidden. When asked why, he was told that he could be taking photographs to be used in promoting a business, said advertisement for which the author might (hypothetically) overlay the picture of a nude woman - thus creating embarrassment for the DNR. The reader is encouraged to look for any such picture in this document.*

Billings General Hospital Base POW Camp
Fort Benjamin Harrison, Indiana

Type Facility	Base POW Camp
Period	February 1944 to November 1945 - Italian POWs (intermittent)
	January 1944 through February 1945 - German POWs (245 days)
Location	Fort Benjamin Harrison (Indianapolis, Indiana)
Num. Prisoners-Avg / Max	11 German PW average over 690 days / 18 peak in December 1944
	5 Italian POW average over 420 days / 12 peak in June 1944
Size / Capacity	Capacity Not Known
MP / Service Units	Not Know (but probably the same unit as Fort Benjamin Harrison)
Branches	None

Indiana Ordnance Works
Charlestown, Indiana

Fort Information regarding this facility, refer to Kentucky: Fort Knox.

Jeffersonville Quartermaster Depot Base POW Camp
Jeffersonville, Indiana

Fort Information regarding this facility, refer to Kentucky: Fort Knox.

Kentucky

U.S. Army Installations – World War II: Fifth Service Command

Kentucky

Army Ground Forces

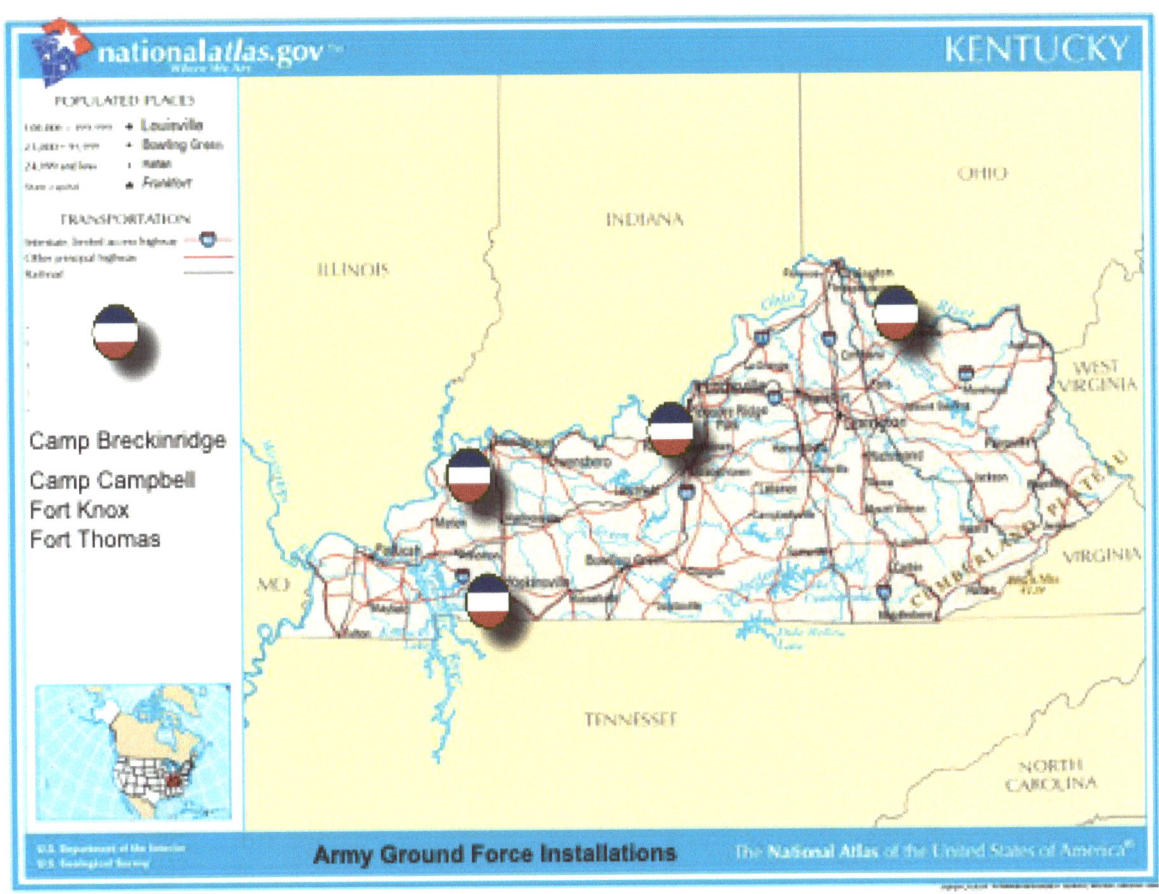

Camp Breckinridge
Camp Campbell
Fort Knox
Fort Thomas

Army Ground Force Installations

Camp Breckinridge

Location	Morganfield, Kentucky *(along U.S. 60 e/o Morganfield)*
Established	1 July 1942 - activated
Inactivated	1 January 1954 - put on inactive status
Size of Facility	50,700 Acres (another source cites 36,000 Acres)
Role	Established as an infantry division training facility. The major units that trained here were the: 35th, 75th, 83rd, 92nd, and the 98th Infantry Divisions. The post also contained a POW camp.
Capacity	2,017 Officers / 41,571 Enlisted personnel 44,133 overall troop capacity

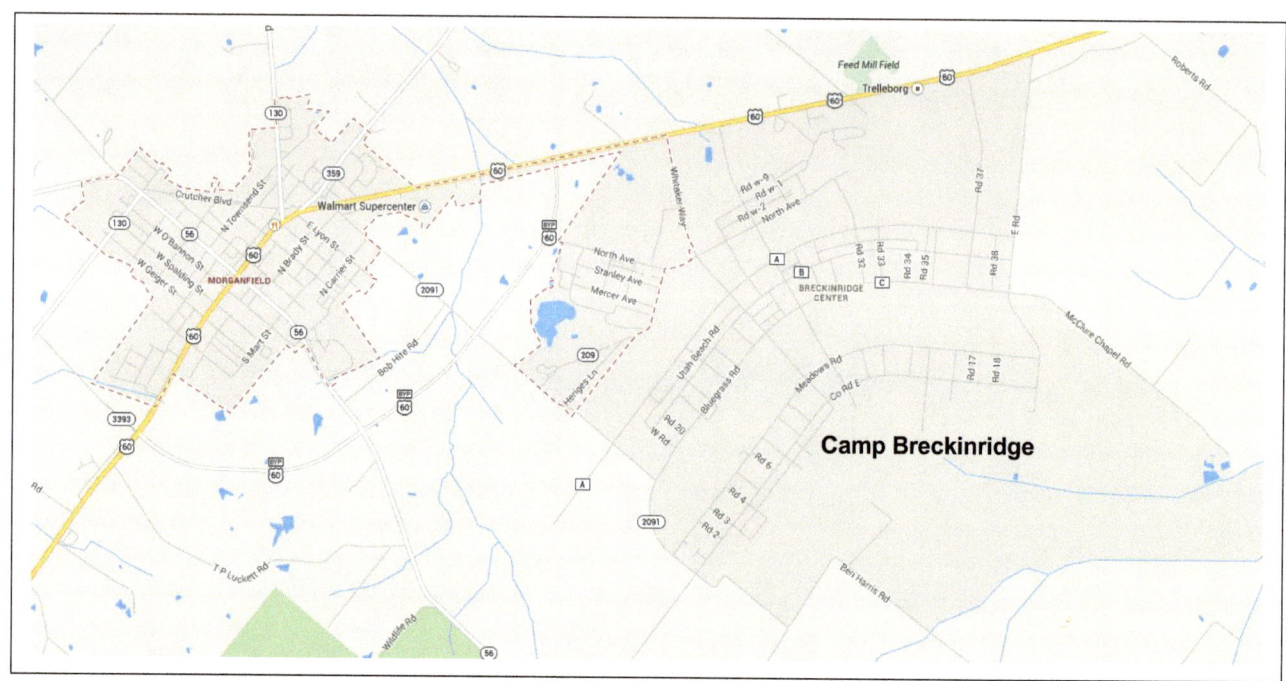

Camp Breckinridge schematic

An oblique aerial photo of Camp Breckinridge (to match to the schematic) (the reader is asked to forgive the quality as the picture is doubtless the product of multiple versions of copy.

(Left) Former officer mess hall; (Right) one of the murals painted by German POWs found in the mess hall. (at center of picture) is a photographic glare – the German artist's work was impeccable; my camera work . . . not so much.

Camp Campbell

Location	Clarksville, Tennessee
	36 38 36.85N 87 26 27.18W - the Main Gate
Established	1942
Inactivated	Remains an operational post today as Fort Campbell
Size of Facility	102,418 Acres (then); 105,068 Acres (now)
Role	This was an armored division camp - the training facility for the 12th Armored Division, the 14th Armored Division and the 20th Armored Division. It was also headquarters for the IV Armored Corps and the 26th Infantry Division. The post also contained a POW camp built to contain 3,000 Germans - Nazis and anti-Nazis.
Capacity	2,449 Officers / 41,687 Enlisted personnel
Notes	The facility spans a vast tract across both Kentucky and Tennessee. The majority of land occupied is in Tennessee, however, the base headquarters is in Kentucky as well as the main gate.

An anomalous condition exists with regard to the bifurcation of the post between two states. It is acknowledged to "reside" at Clarksville, Tennessee, even though the HQ and the main gate are in Kentucky. Statistically, the post is documented in the Station List of the Army of the United States as a camp in Kentucky, in the Fifth Service Command, yet, within this definition the Clarksville, Tennessee, identification is provided - which would be in the Fourth Service Command. **On 6 March 1942 the War**

Department redesignated the address to Tennessee due to proximity to Clarksville.- yet statistically, Camp Campbell remained in the Fifth Service Command, not the Fourth.

An old postcard depicting the officers' quarters and mess halls.

Today's Main Gate

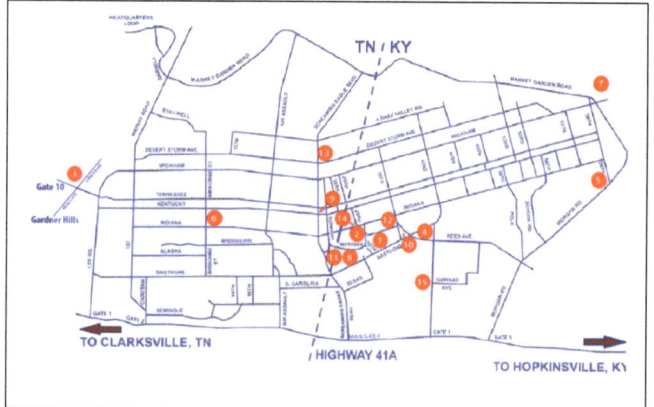

Illustration of how the post is bifurcated by the state line - dividing the cantonment in half.

The main gate in a photo believed to be circa 1950 - not WWII but closer to the WWII reality than the author's current photography.

Fort Knox

Location	Fort Knox, Kentucky
	37 52 59.52N 85 57 55.05 - at the bullion reserve
Established	(staggered history):
	1861 - fortifications built along the river during the Civil War
	1903 - military maneuvers held in the area
	1918 - field artillery training on 20,000 areas near Sitthton, forerunner to permanence
	1922 - became a training area for the Citizens Military Training Camp and the NG.
	1925 - 1928 after adding 40,000 acres, budget downturn stymies camp development the fledgling army post reverts to "Camp Henry Knox National Forest."
	1932 - January 1: the camp is made permanent as it became the newly-established base for mechanized cavalry
	1940 - designated Armored Force Headquarters; later the same year it is designated as the Armored Force School and the Armored Force Replacement Center.
	1950 - upgraded in April to permanent post
Inactivated	Remains an operational post today
Size of Facility	30,346 Acres (then); 109,000 Acres (now)
Role	Headquarters, I Armored Corps, the 1st Armored Division, the Armored Force Board and the Armored Force School. By 1943, Fort Knox had become the country's foremost armored training center. The 1st, 2nd, and 5th Armored Divisions trained here as did dozens of other armored units.
Capacity	3,496 Officers / 53,987 Enlisted personnel; troop capacity - 60,537
Notes	In 1936 the Treasury Department located the U.S. Bullion Depository at Fort Knox. The majority of the nation's gold supply is stored at the facility.
	The post also had a POW camp holding as many as 2,500 prisoners.
	The Patton Museum is on the grounds of the Fort, immediately off the highway. It is an excellent resource for the study of WWII and General Patton.

Display of unit patches of the Third Army - under General Patton

Fort Thomas

Location	Fort Thomas, Kentucky
	39 04 03.22N 84 26 84.08W - at the water tower
	[the post is "sandwiched" between I-471 and the Ohio River]
Established	15 August 1890
Inactivated	One reference cites Fort Thomas as being converted in 1946 to a Veteran's Administration facility, while another source points out that the post was using following WWII until 1964 with a 161 year life span for the military occupation of the facility.
Size of Facility	111 Acres
Role	Fort Thomas served as an Induction Center, Replacement Depot, and Military Hospital.
Capacity	3,433
Notes	The pre-eminent fixture on the post is the water tower which was built to commemorate involvement in the Spanish-American War. Over the years various infantry units had garrisoned at Fort Thomas including the 2nd, 3rd, 4th, 9th, and 10th Infantry Regiments.
	Fort Thomas was the "home post" for Sergeant Samuel Woodfill, the most highly ecorated soldier of World War I (like Audie Murphy for World War II).

Officer Quarters (circa 1900)

*The Water Tower
(left) 1900; (right) 2012*

Officer Quarters 2012

Historic "Reminder"

Army Air Force

Anchorage Field

Location	Louisville, Kentucky *(about 2.5 miles southeast of central Bowling Green)*
Established	Not known
Inactivated	Not known
Units	Not known
Notes	Anchorage was a satellite field to Bowman Field and operated for appr. 6 months.
Auxiliaries	None

Bowling Green Municipal Airport

Location	Bowling Green, Kentucky *(14 miles southeast of Camp Atterbury)*
Established	1934 - as a military airport

Inactivated	Indications are that the airport was used for several months in 1943 and for several in 1944. Records suggest that the airport was turned over to the public domain afterwards.
Units	Not known
Auxiliaries	None

Bowman Field

Location	Louisville, Kentucky *(5 miles southeast of downtown Louisville)*
Established	1921 - Bowman had AAF units in place as early as mid-1941. Opened for the war effort in April 1942 with the I Troop Carrier Command; June 1942 - activated
Inactivated	The date is indefinite, but most likely was tied to the end of the war.
Role	The primary function of Bowman Field was to serve the I Troop Carrier Command which used the air base for air evacuation units and glider pilots. There was also a school of aviation medicine and a Headquarters for the Persosnnel Distribution Command.

I Troop Carrier Command; Glider Pilots Combat Training units; Air Medical Evacuation units. The I Troop Carrier Command was based at Stout Field, Indianapolis. The role of this unit was to fly supplies, troops and nurses to the front line and take out the wounded and sick troops. |
| **Size of Facility** | The field was 426 Acres; troop capacity - 4,670 |
| **Units** | 27th Air Base Unit
28th Air Base Group |
| **Notes** | During World War II, Bowman Field was one of the nation's most important training bases as well as the nation's businest airport. The facility became known as Air Base City when a bomber squadron moved in and more than 1,600 recruits underwent basic training in a three-month period. The Army Air Force's school for flight surgeons, medical technicians, and flight nurses also called Bowman Field home. [1]
Bowman Field is Kentucky's first commercial airport and is the oldest continually operating commercial airfield in North America.

As of 1938, the field was known as "Bowman's cow pasture" because it ws mainly a barren field yet to be developed and modernized. In 1943 and 1944 the average number of men and women serving at the field was 3,277 (2/3 men, 1/3 women). |
| **Auxiliaries** | Anchorage Field
Lexington Army Air Field
Milroy Auxiliary Field (Milroy, Indiana)
Zenas Auxiliary Field (Zenas, Indiana) |

Bowman Field layout – during the war and recently

Bowman Field Headquarters "then" and "Now" with an illustration of part of the edifice depicting the Army Air Corps symbol. Each concrete inlay contains one of these symbols.

Camp Breckinridge Army Air Field

Location	Camp Breckinridge, (Morganfield), Kentucky
	37 41 31N 87 52 36W
Established	1943
Inactivated	1946
Mission	The field was used as a training camp for infantry units up to the division in size.[17] More likely this was used in support of base operations for Camp Breckinridge.
Units	Not known
Notes	Military operations ceased at the close of the war but the field had a continuing life. A number of closure dates have been speculated, therefore, the actual date is not known
Auxiliaries	None

Godman Field

Location	Fort Knox, Kentucky *(located on the army post)*
	37 54 15N 85 57 56W
Established	1937
Inactivated	Reassigned to Fourth Air Force in 1946; remains operational, but under the U.S. Air Force
Mission	1. Support army force operations at Fort Knox
	2. With the 73rd Observation Group - train and provide tactical reconnaissance and support for any army maneuver.
	3. First Air Force training station
Capacity	2,860 troops
Units	99th Army Air Force Base Unit
Notes	Following the war, the field became the Godman Air Force Base from 1947-1954.
Auxiliaries	Milport Auxiliary Field (Milport, Indiana)
	St. Anne Auxiliary Field (North Vernon, Indiana)

Camp Campbell Army Air Field

Location	Hopkinsville, Kentucky *(located at the north end of Camp Campbell)*
	36 39 58N 87 29 06W
Established	1941
Size of Facility	Records hint at the size of the air field as being on 170 acres; however, today, the size of the field is 2,500 acres making it the largest army air field in the country.
Role:	Given the size of the field during the war, as with Schoen Field, Godman Field, and numerous others around the country located on an army ground force post, the role was most likely in support of that post's operations.
Auxiliaries	Campbell had no auxiliary fields of its own, however, is served as a sub-base to Godman Field, to William Northern Field, and to Smyrna Army Air Field - for the longest period.
Units:	99th Base Headquarters & Air Base Squadron, Detachment

Footnotes   17   Sturgis Army Air Field was the primary field for use by Camp Breckinridge, therefore, the observation made by Osborne about this being a training field for Breckinridge comes into question. More likely, the field was used as in support of operations of the army post as with many of the post-based airfields of the era.

The Tennessee - Kentucky state line divides Camp Campbell such that approximately 1/2 of the camp is located in each state. At one point during WWII, the headquarters was transplanted into the Tennessee section, however, all data such as through the *Station List and the Directory of the Army of the United States* the camp is registered in the Fifth Service Command (Kentucky). Tennessee falls into the Fourth Service Command.

In the case of the Campbell Army Air Field, a part of the base structure, it is located on the Kentucky side and can be seen in the aerial photo above. The approximate location of the state line is superimposed.

Goodall Field

Location   Danville, Kentucky   *(less than 5 miles south of downtown Danville)*
Established   No details available for this installation
Size of Facility   Currently the field is reported to be 170 acres; however, the WWII size is not known.

Kenton County Airport

Location   Covington, Kentucky
Established   12 August 1944 - opened; October 1942 funds approved for construction
Inactivated   Declared surplus on 27 October 1946

Units	The airport served under the Air Ferrying Command
Size	928 Acres
Notes	Due to frequent fog conditions at the primary Cincinnati area airport, Lunken Airport, the case was made to build another facility in the greater Cincinnati area. The facility was approved in October 1942 and the first B-17 first used the field shortly after its official opening on 12 August 1944. Another key consideration for seeking a new airport. . . on higher ground in Kentucky, was to avoid Ohio River flooding.
Auxiliaries	None However, it was an auxiliary to Lockbourne Army Air Field (Columbus, Ohio)

Lexington Municipal Airport   (aka Bluegrass Airport)

Location	Lexington, Kentucky *(a little over 5 miles due west of downtown Lexington)* 38 02 11N 84 36 21W
Established	Not known
Inactivated	Not known
Units	Not known
Notes	Lexington Municipal Airport was an auxiliary to Bowman Field. As an auxiliary to Bowman, it would have come under the control of the Troop Carrier Command.
Auxiliaries	None

Paducah Municipal Airport   (aka Barkley Regional Airport)

Location	Paducah, Kentucky *(a little under 10 miles w-southwest of downtown Paducah)* 37 03 37N 88 46 23W
Established	Not known
Inactivated	Not known
Units	Not known
Notes	Not known
Auxiliaries	None

Standiford Field

Location	Louisville, Kentucky *(located along I-65 between I-264 and I-265)*
	38 11 09.43N 85 44 31.08W - at terminal
Established	1941 - built by the Corps of Engineers
Inactivated	1947 - the field was turned over to the public
Role	The field was used by the Air Technical Service Command along with the development of a large manufacturing plant for aircraft for the C-76 transport.
Units	10th Reconnaissance Group - 7 Nov 1941 to 23 June 1943
	387th Bombardment Group - 11 May 1943 to 10 June 1943
	391st Bombardment Group - 4 Sept 1943 to 31 Dec 1943
	In
Notes	The air field was originally known as Standiford Field but later the name was changed to Louisville Municipal Airport and then once again to Louisville International Airport.
	Bowman Field had been one of the busiest airports in the nation prior to the opening of Standiford Field. When opened, most operations were removed from Bowman and centered at Standiford Field.
	As with Kenton County Airport (Cincinnati International Airport), Standiford Field was built at the given location on ground to known to not have flooded.
Auxiliaries	None

Sturgis Army Air Field   (aka Sturgis Municipal Airport)

Location	Sturgis, Kentucky *(1.5 miles east-southeast of downtown Sturgis)*
	37 32 26.25N 87 57 27.08W
Established	21 May 1943
Inactivated	Late-1946 - the base was declared surplus
Size of Facility	307 Acres; troop capacity - 450
Units	506th Parachute Infantry Regiment - 6 June 1943 to 23 July 1944
Notes	The operation of the field was taken over by the I Troop Carrier Command and used as a sub-base of George Field (IL) for training purposes.

U.S. Army Installations – World War II: Fifth Service Command

Army Service Forces

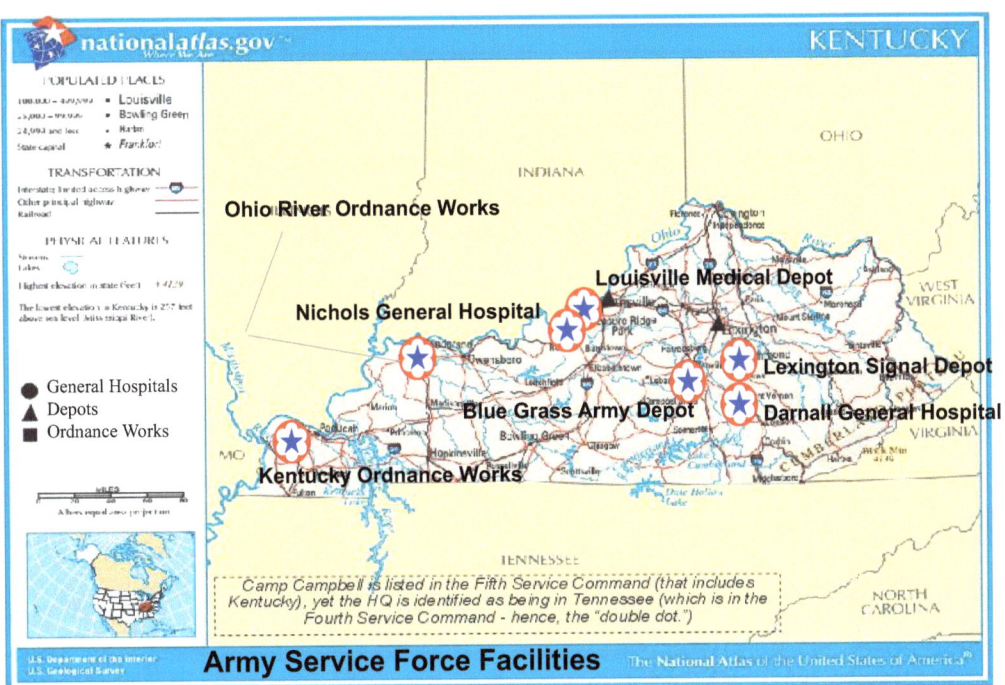

General Hospitals

Darnall General Hospital

Location	Danville, Kentucky (vicinity)
	37 42 27N 84 44 46W
Established	1941
Inactivated	1946
Size of Facility	551 Acres
Troop Capacity	513
Role	The facility was initially constructed as a state mental hospital, however, beginning in 1941 the Army assumed control of the facilities to provide care for soldiers suffering from psychiatric illness. Darnall also provided an internment facility for German POWs.
Notes	The facility became the Kentucky State Hospital between 1946 and 1977 at which time control of the facility was relinquished and turned over to the Bureau of Social Services. From 1977 to 1983, the Danville Youth Development Center utilized the facility.

In January of 1983 the Kentucky Department of Corrections gained control of the property and converted the installation into the Northpoint Training Facility - a minimum security institution for fewer than 500 inmates. Over the years the facility's program has changed and it is now a "medium security institution" with 1,256 inmates.

In some cases, the name is found to be misspelled as Darnell.

Nichols General Hospital

Location Louisville, Kentucky
38 11 18N 85 47 53W
Established November 1942
Inactivated 31 March 1946 - turned over to the Veterans Administration
Size of Facility 1,000-Bed Hospital
Troop Capacity 1,510
Role General hospital care for the *sick and wounded soldiers* - a "temporary facility." The new VA facility became the Nichols VA Hospital and operated by the VA until 1952.

Then

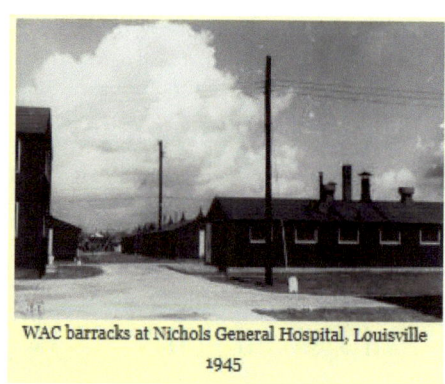

WAC barracks at Nichols General Hospital, Louisville 1945

Now

Today, the area that was home to the Nichols General Hospital, is a mix of single-family homes and multi-family homes in a marginal part of town. During the war, this was beyond the city limits but today is within the city, just north of I-264 and west of the Churchill Downs.

Ordnance Plants & Works

Kentucky Ordnance Works

McCracken County, Kentucky   (outside of Paducah, KY)
37 06 38N 84 48 49W

In December 1942 the KOW site was established west of Paducah for the manufacturing of explosives. The facility was 16,126 acres and produced 196,490 tons of TNT. Operations were shut down in August 1945. Initially, the Atomic Energy Commission received the site but the land is now owned between the Tennessee Valley Authority and the West Kentucky Wildlife Management area.

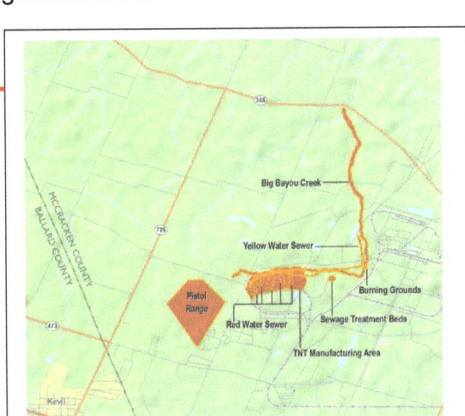

Ohio River Ordnance Works

Henderson, Kentucky

The ordnance facility was built in 1941 - records are not readily obtainable.

Depots

Blue Grass Army Depot

Richmond, Kentucky
37 40 46 N 84 15 18W

The plant which was established in 1941 and began operations in 1942 as an ammunition and geeral supply storage depot, closed in 1999 as part of the Base Realignment and Closure program. The facility was merged in 1964 with the Lexington Signal Depot in Avon, KY to become the Lexington-Blue Grass Army Depot.

The facility contains 14,594 acres with 1,153 buildings, 902 igloos and storage capacity of 3,233,598 square feet. The base is mostly open fields and wooded land and is used for munitions storage, repair of general supplies, and the disposal of munitions. [18]

Footnotes   (18) http://en.wikipedia.org/wiki/Blue_Grass_Army_Depot

Entrance to what is now the Blue Grass Chemical Activity - responsible for the chemical weapons stored at BGAD - part of the Army's Chemical Materiels Acity headquartered at Edgewood, Maryland.

Lexington Signal Depot

Lexington, Kentucky
38 04 18N 84 18 55W

The Army began construction of what is now known as Bluegrass Station July 1, 1941. It was originally designated as the Lexington Signal Depot and later called the Lexington Army Depot. The post was established by War Department General Order No. 6, dated June 25, 1941.

The depot was originally built on 782 acres of Central Kentucky farmland, located in the eastern section of Fayette County. The base soon grew to include the administration building, eight 130,000 sq. ft. brick warehouses, the motor pool, power plant, and 40 temporary structures. Many additions have been made since the original construction, including two 200' X 1200' concrete block warehouses, a heliport, two hangars, and a large maintenance building.

The mission of the depot consisted of a variety of duties, including storage and shipment of supplies and ammunition, testing of radio equipment, and development of new technologies. Those stationed at the Lexington Army Depot made many valuable contributions to the war effort through these duties.

During World War II, as it was with most of the nation, women played an important role in the depot's wartime effort. Their duties ranged from forklift operation and maintenance to guarding the installation.

Approximately 35 German Prisoners of War (POWs) were brought to the depot in February 1945 to construct barracks to be used to house other POWs that were to follow. Approximately 275 German prisoners were held in these barracks. During this time they were used to process unserviceable wire, load and unload supplies, and maintain roads and grounds. The prisoners were moved to Ft. Knox, Kentucky, in February 1946.

After World War II the base mission was expanded to include use as a storage depot for materials and supplies such as dry cell batteries, clothing, textiles, and various manufactured metal products. During the 1950's and 60's, the depot continued to provide support for our troops overseas in the Korean Conflict and the Vietnam War. During this time, the depot participated in the research and development of new technologies that were used throughout the military, in addition to its regular duties.

The Lexington Army Depot merged with the Bluegrass Depot in Richmond, Kentucky, in August 1964, thus creating the Lexington-Bluegrass Army Depot (LBAD). It remained the LBAD until December 1988, when the Department of Defense Base Realignment and Closure (BRAC) Commission announced the closure of the Lexington Facility. [19] The facility had a troop capacity of 364.

Footnotes   (19) http://kynghistory.ky.gov/places/bluegrassstn.htm

Louisville Medical Depot
Louisville, Kentucky

The facility served as a reserve medical supply depot; it stored and issued medical supplies for International Aid (lend-lease). The facility performed a range of fuctions: holding and reconsignment point (1942), war aid depot (1942), quartermaster depot (September 1942), and ordnance depot (10 October 1942). [20]

The facility was activated as the Louisville Medical Depot on 24 July 1943; its warehousing capacity was 1, 713,000 square feet in size.

At the outset of World War II, the Medical Department was responsible for procurement, storage, and issue of approximately 4,500 items. By 1942, the number of items under medical cognizance had jumped to 6,000, and another 1,000 items were added by 1943. During mid-1942, some 700 contractors were serving the Medical Department, but within a year's time, this had jumped to 2,500 contractors and the number of contracts had reached 25,000.

Aggregating more than 5.4 million square feet by July 1941, Medical Supply Depot space continued to expand after entry of the United States into the war. As more Medical Depots were built in order to accommodate the introduction of the 7,000 additional items required, more depots were opened in July 1942, making the total space available 7 million square feet.

Additional space acquisitions increased storage facilities still further during the next 6 months; in July 1943, the peak of 13 million square feet was reached. From this time until the end of the war, the amount of space available was steadily dropping. This was partly as a result of the demand for storage area diminishing. Twenty Depots and Medical Sections had been occupied in July 1943. This number dropped to 17 in July 1944, and to 14 in July 1945. During the same period, the number of square feet occupied fell to 10,348,000 in 1944, and to 9,127,000 in 1945. [21]

Footnotes   (20) history.armedd.army.mil/bookdocs/wwii/medicalsupply/appendixa.pdf
            (21) www.med-dept.com/articles/medical-depots-in-the-zone-of-the-interior/

Generic photo of a medical depot in the Zone of the Interior

U.S. Army Installations – World War II: Fifth Service Command

Army Service Force Locations

Location	Branch	Type Facility
Ashland	ASF Schools	Ashland High School
	Districts	Armed Forces Induction Districts
Bowling Green	Districts	Armed Forces Induction Districts
	ASF School	ROTC Schools
Danville	Army Service Forces	General Hospitals
Fort Knox	Army Service Forces	Rehabilitation Center
	Boards	Boards
	Centers	Replacement Training Centers
	Field Printing Plant	Field Printing Plant
	Laboratories	Medical Research Laboratory
	Schools	Schools
		Officer Candidate School
		Officer Replacement Training Center
		Quartermaster Schools
Fort Thomas	Army Service Forces	Reception Center
Frankfort	Districts	Armed Forces Induction Districts
Henderson	Depots	Plants and Works
Hopkinsville	Army Service Forces	Prisoner of War Camps
		Laboratories (Post Photographic)
		Ordnance Service Command Shops
	Schools	Quartermaster Schools
Lexington	Schools	Signal Corps Schools
		University of Kentucky
	ASF Schools	ROTC Schools
	Army Service Forces	Laboratories (Post Photographic)
	Depots	Signal Depots
	Pools	Signal Corps Officer Replacement Pool
	Remount Areas	Remount Areas
	Districts	Armed Forces Induction Districts
Louisville	Army Service Forces	Army Reservation Bureau
		General Hospitals
	Schools	Medical Schools
		Civilian Schools
	ASF Schools	ROTC Schools
	Depots	Medical Depots
	Quartermaster	Quartermaster Market Centers
	District	District Engineers
	Districts	Armed Forces Induction Districts
	Schools	[Bowman Field] Quartermaster Schools
Lyndon	ROTC Schools	ROTC Schools

Army Service Force Locations

Location	Branch	Type Facility
Middleboro	Districts	Armed Forces Induction Districts
Morganfield	Army Service Forces	Prisoner of War Camps
		Laboratories (Post Photographic)
		Ordnance Service Command Shops
	Schools	Quartermaster Schools
Owensboro	Districts	Armed Forces Induction Districts
	ASF Schools	ROTC Schools
Paducah	Districts	Armed Forces Induction Districts
	Depots	Plants and Works
Pikesville	Districts	Armed Forces Induction Districts
Richmond	Schools	Civilian Schools
	Schools	Administrative Schools
		Eastern Kentucky State Teachers College
	ASF Schools	ROTC Schools
	Depots	Quartermaster Depot

Prisoner of War Camps

The state of Kentucky had three base POW camps at Fort Knox, Camp Campbell, and at Camp Breckinridge. In the case of the facility at Camp Breckinridge, it served as both a base camp for a period and then as a branch to Camp Campbell. Another consideration was that there were three branch camps located in the state of Indiana - at Jeffersonville and at Charlestown - both adjacent to Louisville on the north side of the Ohio River.

Fort Knox Base POW Camp
Fort Knox, Kentucky

Type Facility	Base POW Camp
Period	Italians were held at Fort Knox between March and May, 1944, whereas German prisoners were held between June 1944 and March 1946
Location	Located at the Fort Knox post
Size / Capacity	Capacity varied: 750 / 2,101 / 2,100 / 2,015
Num. Prisoners-Avg / Max	3,098 German POW average over 660 days / peak of 5,036 in July 1945
	657 Italian POW average over 90 days / 743 peak in March 1944
Nationality	Italian (first); replaced by Germans
MP / Service Units	414th Military Police Escort Guard Company (ASF)
	3571st Service Command Unit: Prisoner of War Camp
Branches	**Bowman Field Branch Camp**, Louisville, KY
	The air field branch held German POWs from December 1945 through March 1946.
	155 average over 120 days / 160 peak in January 1946
	Charlestown Branch Camp, Charlestown, IN
	[See Indiana POW Camps]
	Danville Branch Camp, Danville, KY
	German prisoners were held here in April and May 1945 and then again in September and October 1945.
	151 average over 120 days / 242 peak in September 1945
	Darnall General Hospital Branch Camp, Danville, KY
	Operated for 61 days holding a total of 630 German prisoners between September and November 1945
	150 average over 180 days / 152 peak in September 1945
	Eminence Branch Camp, Eminence, KY
	German POWs were held at the facility in Eminence during October 1944
	316 average over 30 days / 316 peak in October 1944

Frankfort Branch Camp, Frankfort, KY
German prisoners were held at Frankfor between October 1944 and October 1945 in non-consecutive months: October 1944 and then again from September to October 1945.
335 average over 90 days / 526 peak in September 1945

Jeffersonville Branch Camp, Jeffersonville, IN
[See Indiana POW Camps]

Lexington Branch Camp, Lexington, KY
German prisoners were held here in March 1944 and again beetween March and October 1945.
307 average over 270 days / 601 peak in October 1944

Lexington Signal Depot Branch Camp, Lexington, KY
From July 1943 through May 1944 (300 days) the facility held Italian POWs and from June 1944 through June 1946 (730 days) it held German POWs.
209 average over 180 days / 242 peak in September 1945

Louisville Branch Camp, Louisville, KY *(location not known)*
German prisoners were held here between May and November 1945
196 average over 210 days / 232 peak in September 1945

Maysville Branch Camp, Maysville, KY
This facility also held German POWs - briefly, during October 1944.
300 average over 30 days / 300 peak in October 1944

Paris Branch Camp, Paris, KY
This facility operated in September and October 1945 for German POWs
423 average over 60 days / 575 peak in September 1945

Shelbyville Branch Camp, Shelbyville, KY
This suburban Louisville facility held German POWs between July and October 1945.
308 average over 120 days / 328 peak in August 1945

Camp Breckinridge Base POW Camp
Morganfield, Kentucky

Type Facility	Base POW Camp; subsequently a Branch Camp of Camp Campbell
Period	The camp operated between July 1943 and February 1946 (946 days) for German POWs.
Location	Located at the Camp Breckinridge post
Size / Capacity	Capacity varied: 3,000 / 3,060 / 3,000
Num. Prisoners-Avg / Max	1,748 average over 960 days / 2,730 peak in August 1944
Nationality	German POWs

MP / Service Units	405th Military Police Escort Guard Company (ASF)
	406th Military Police Escort Guard Company (ASF)
	646th Military Police Escort Guard Company (ASF)
	1538th Service Command Unit: Prisoner of War Camp
Branches	Owensboro Branch Camp, Owensboro, KY
	Owensboro Branch operated from July 1945 through October 1945 holding German POWs
	257 average over 120 days / 424 peak in September 1945

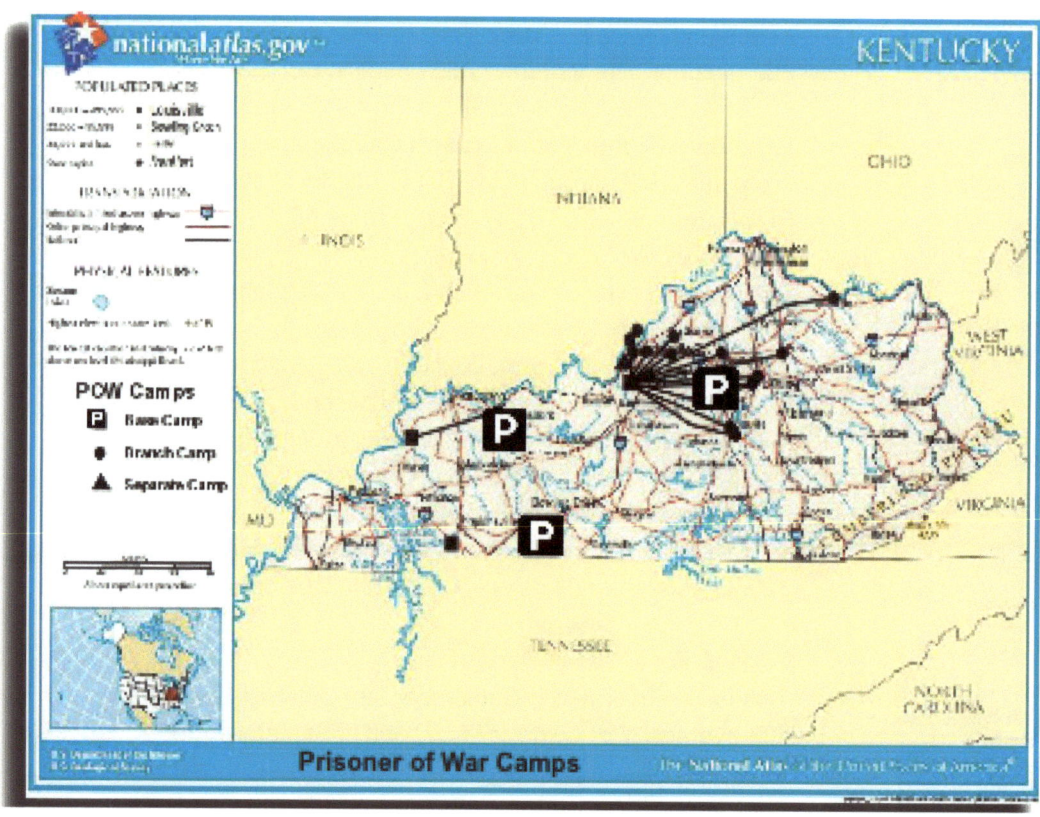

Camp Campbell Base POW Camp

Hopkinsville, Kentucky

Type of Facility	Base POW Camp
Period	The camp operated for german POWs between August 1943 and May 1946 (1,004 days)
Location	Located on the post at Camp Camplell
Size / Capacity	Capacity varied: 3,000 / 3,111 / 3,110 / 3,000
Num. Prisoners-Avg / Max	1,065 average over 1,020 days / 2,349 peak in August 1944
Nationality	German POWs
MP / Service Units	494th Military Police Escort Guard Company (ASF)
	658th Military Police Escort Guard Company (ASF)

U.S. Army Installations – World War II: Fifth Service Command

	1539th Service Command Unit: Prisoner of War Camp
Branches	None

Nichols General Hospital Separate POW Camp

Louisville, Kentucky

Type Facility	Penal / Medical Institution (PMI) POW Camp
Period	The camp operated from November 1944 through January 1945
Location	Located at Nichols General Hospital (southwest Louisville)
Size / Capacity	Limited capacity
Num. Prisoners-Avg / Max	2 average over 90 days / 2 peak in November 1944
Nationality	German POWs
Branches	None

69

Ohio

Ohio

Army Ground Forces

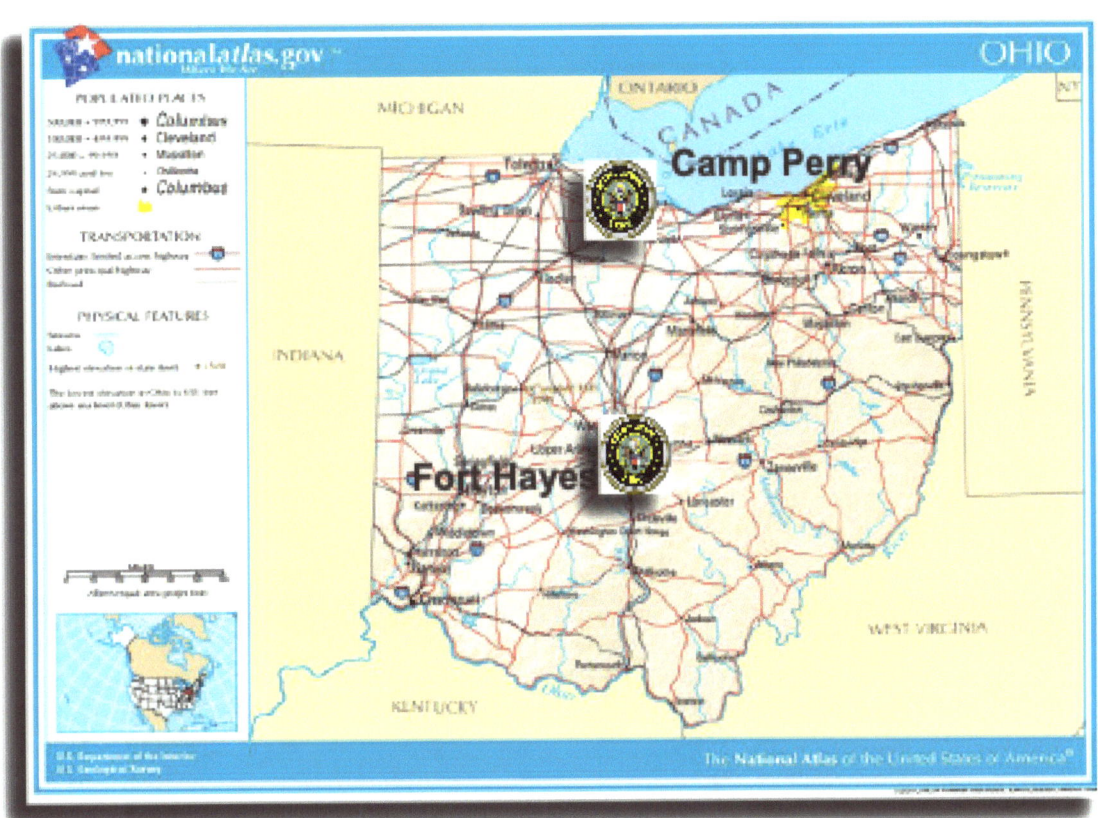

Fort Hayes

Location	Columbus, Ohio *(immediately north-northeast of downtown Columbus)* 39 58 27N 82 59 15W
Established	17 February 1863
Inactivated	30 June 1947 - the post was determined to be surplus; then abandoned
Size of Facility	70 Acres *(77.5 actual)*
Role	Fort Hayes served as the administrative headquarters for the Fifth Service Command during the war and the HQ for the Fifth Corps Area prior to the renaming of the districts in the country to the various service commands. The post served as a Reception Center during the war beginning in 1940 with more than 200,000 men processed into the service through the post. This was deactivated on 1 March 1944.
Notes	Fort Hayes was originally known as the Columbus Barracks on land deeded to the Army - this remained until after World War I when on 13 Dec 1933 it became Fort Hayes.

Today, 21 acres have been retained for the U.S. Army Reserve and 50 acres have been deeded over to the Columbus School system.

Of note - the most prominent building on the post is known as the "shot tower." In the tower molten lead was dropped in beads from the heights of the tower and the natural forces of gravity formed the lead into balls, or shot and in musket balls. The facility was operated by the Ordnance Department.

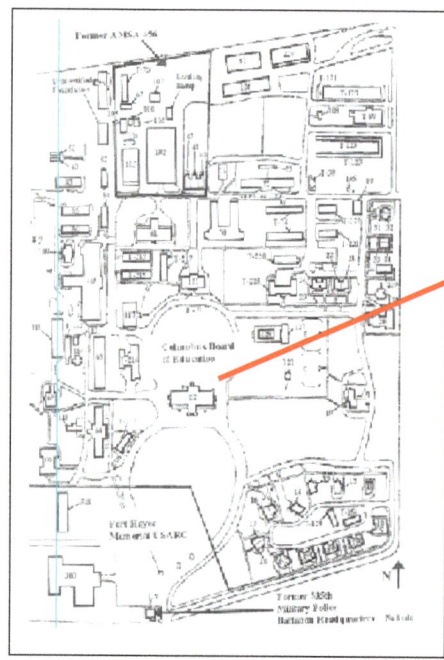

Several examples of the state of buildings at Fort Hayes.

The majority of the building, not assigned to the Columbus School system grounds, are largely rundown and falling apart. Most have a character of World War I facilities or even older.

Camp Perry [22]

Location	LaCarne, Ohio (approximately 18 miles west-northwest of Sandusky, OH)
	41 31 51N 83 01 12W
Established	1906 - activated
Inactivated	15 January 1946 - declared surplus
Size of Facility	640 Acres (originally 521 acres, expanded to 1,500 acres and back to 640)
Role	Camp Perry served as a Reception Center for the Fifth Corps Area
	As of 30 September 1943, Camp Perry was designed as a base POW camp until such time that the prisoners were returned abroad on 13 December 1945.

U.S. Army Installations – World War II: Fifth Service Command

Camp Perry had been a National Guard training camp but was used as a POW camp during the war.

A lack of soldiers with marksmanship skills in the Spanish-American War led to the founding of Camp Perry 1906. The camp grew to be one of the largest rifle ranges in the U.S. It was used to train soldiers during World War I and World War II. The camp also housed German and Italian prisoners of war. Later, Camp Perry became a training facility for the Ohio National Guard. The post is reputed to have the best rifle and pistol range in the U.S. and many national competitions - military and non-military, are held here. Numerous huts used by the POWs are still in place at the camp.

Notes
Footnotes (22) *Sources:*

http://www.globalsecurity.org/military/facility/camp-perry.htm
http://www.wipipedia.org/wki/Camp_Perry
http://www.oll.state.oh.us/your_state/remarkable_ohio/arker_details.cfm?marker_id=119&file_id=4672

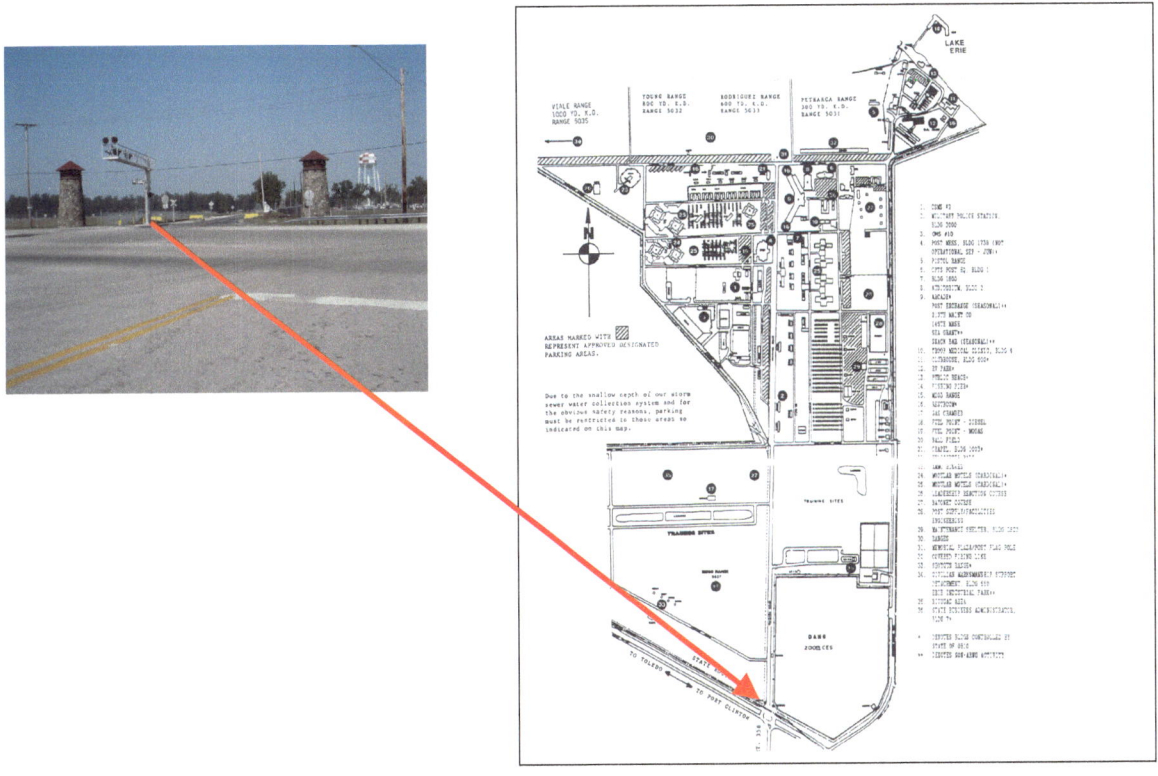

Camp S

Location)ughly a quarter mile north of downtown Chillicothe)

Establish
Inactivate
Size 3,417 Acres; 2,000 acres initially created the 3rd largest training camp in the U.S. (at that time)

Role Camp Sherman was not activated during World War II. [23] During World War I, the 83rd Division had been garrisoned here, on a post built atop a former Hopewell Indian burial mound. [23]

Footnotes (23) *Camp Sherman was not activated during World War II but is listed (in brief) here as it was visited and researched along with so many former Army facilities. There was a natural curiosity as to why it wasn't activated.*

(24) *Peck, G. Richard, The Rise and Fall of Camp Sherman, Craftsman Printing Company, Chillicothe, OH, 967.*

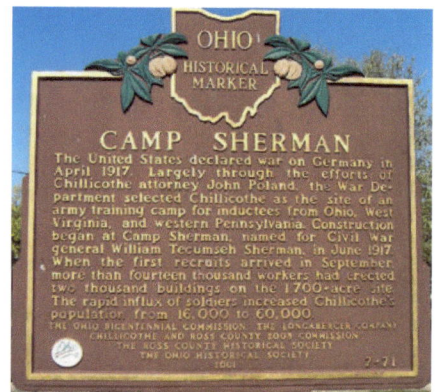

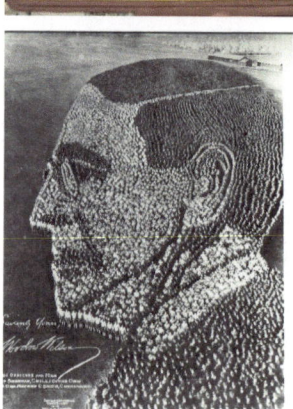

Aerial of Camp Sherman (below) and commemorative plaque at the site.

21,000 soldiers from the 83rd Division formed up in the likeness of President Woodrow Wilson at Camp Sherman.

U.S. Army Installations – World War II: Fifth Service Command

Army Air Force

Cleveland Municipal Airport

Location	Cleveland, Ohio
	41 24 37.98N 81 50 13.07W
Established	1925 - the first municipally-owned airport in the U.S.
Inactivated	Not known
Mission	The Army Air Force used the field as part of its air transport system and as an AAF Flight Service Center.
Units	512th AAF Base Unit (HQ, ATC Field Authorization, Operating Location
	553rd AAF Base Unit (3rd Ferrying Group), Operating Location #9
Notes	Through 1944 the airport was associated with the Cleveland Aircraft Assembly Plant #7 (Fisher Body Division, GMC) and in 1945 as the AAF Flight Service Center. At that point the assembly plant had been redesignated as the Government Owned Assembly Plant #7. The assembly plant was located immediately to the east and south of the current terminal area.
	This is now the regional airport for the Cleveland metropolitan area, the Cleveland Hopkins International Airport.
Auxiliaries	None

*Cleveland Municipal Airport
World War II era*

Clinton County Army Air Field

Location	Wilmington, Ohio 39 25 54N 83 47 22W
Established	1942 - the Army took over operation of the airport
Inactivated	1945 - the facility was closed; reopened during the Korean War; closed again in 1971
Mission	The field was used by the Army Air Force to test gliders.
Notes	The airport opened in 1929 and a small hangar was built in 1930. The landing strip was approved by the Civil Works Administration in 1933. In 1940, the Civil Aeronautics Authority took control of the Wilmington Airport for use as an emergency anding field. In 1942 the Army Air Corps took over the airport renaming it Clinton County Army Air Field. The Air Materiel Command used the airfield for glider research, training and development until the end of World War II. The airfield was closed after World War II, but reopened during the Korean War. By 1958, the Clinton County Air Force Base was home to the newly created 249th Air Reserve Training Wing. The runway was extended from 6,000 feet to 9,000 feet in 1960. The air force base was closed permanently in 1971 and its operations transferred to Lockbourne Air Force Base in Columbus.
Auxiliaries	None

Wilmington Airport /
Clinton County Army Air Field

Dayton Municipal Airport
(aka) Dayton Army Airfield

Location	Vandalia, Ohio 39 53 53N 84 13 22W
Established	December 1941
Inactivated	15 September 1947 - quit claim deed to the City of Dayton
Mission	The Army Air Force used the field as a sub-base to Wright Field.
Notes	Over the several years following the commitment to use the airport, WWII played a significant role in the growth of the Dayton Municipal Airport. During the period of 1942 - 1945, the U.S. Department of Defense (sic), the War Department, acquired various real estate interests in and around the airport for the construction of an army training airfield.

In support of the war efforts, the U.S. Army leased the airport. Pursuant to Army General Order #72 (4 Sep 1944), the airport had been designated the "Dayton Army Airfield."

On September 15, 1947, the Dayton Municipal Airport became the largest commercial airport in Ohio when the federal government, acting through the War Assets Administrator, deeded "Dayton Army airfield" containing over 551 acres of property and related military facilities to the City of Dayton and extinguished its lease of the airport.

Lockbourne Army Air Field

Location	Columbus, Ohio
	39 49 01N 82 56 07W
Established	1942
Inactivated	1991
Mission	This was another field intended for the I Concentration Command converted to the Southwest Training Command and used for training pilot instructors of 4-engine planes. After the war, the flight training activities were halted and the airfield was used as a development and testing facility for all-weather military flight operations.
Notes	Following World War II, Lockbourne AAF went through a number of changes: Northeastern Training Center of the Army Air Corps
	Lockbourne Army Air Base
	Lockbourne Air Force Base
	Rickenbacker Air Force Base
	In 1991, Rickenbacker (named for WWI aviation ace Eddie Rickenbacker) was realigned and began being used by the Ohio Air National Guard.
Key Units	374th Base Headquarters and Air Base Squadron
	2114th AAF Base Unit (Pilot School, Specialized
	4-Engine & Instructor School)
	(refer to appendix for all units)

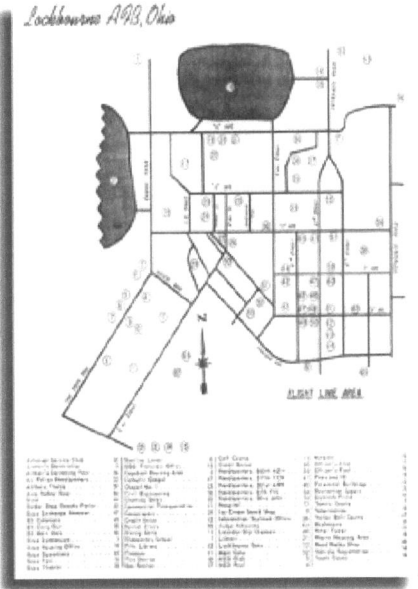

Schematic of Lockbourne AAF

Lunken Airport

Location	Cincinnati, Ohio
	39 06 13.34N 84 25 44.49W
Established	"Early WWII"
Inactivated	Not known
Size of Facility	1,000 Acres
Mission	The field was taken over by the I Concentration Command to become its Headquarters, but was short-lived. It had general military use by various branches.
Notes	Cincinnati Municipal Airport, also known as Lunken Airport, was a commercial airport in the 1920s, 1930s, and into the 1940s. It is situated in the Little Miami River valley near Columbia, the site of the first Cincinnati-area settlement in 1788. When the original 1,000 acre airfield was dedicated in 1925, it was the largest municipal airfield in the world. American Airlines started at Lunken Airport during this time, but no major commercial airlines use it anymore. Instead, many large Cincinnati-area companies now base their corporate aircraft there.
	On December 17, 1925, the Embry-Riddle Company was formed at Lunken Airport by T. Higbee embry and John Paul Riddle. A few years later, the company moved to Florida, and later became the Embry-Riddle Aeronautical University.
	Lunken Airport was supplanted by the Cincinnati / Northern Kentucky International Airport in 1937 following catastrophic flooding from the Ohio River. The flooding problem prompted the airport's nickname of "Sunken Lunken." The control tower, located at the southwest corner of the airport, was almost totally submerged during the historic Ohio River floof of 1937, and now has a single black brick facing the airfield to indicate the high - water mark. Today the old control tower is home to the Lunken Cadet Squadron of the Civil Air Patrol, and is the oldest standing control tower in the U.S. The property also contains public recreation areas, including an 18-hole golf course, playgrounds, and walking / biking paths. [25]
Key Units	586th AAF Base Unit (Ferrying Squadron, ATC)
	502nd AAF Base Unit (HQ, ATC Field Authorization), Operating Location
	551st AAF Base Unit (Ferrying Division, AFATC, Reserve)
	(for complete list refer to appendix)
Footnotes	(25) http://en.wikipedia.org/wiki/Cincinnati_Municipal_Lunken_Airport

Patterson Field

Location	Dayton, Ohio
	39 49 41N 84 02 44W
Established	During World War II
Inactivated	Realigned with Wright Field to form the Wright-Patterson Air Force Base
Size of Facility	Not known
Mission	Although this was used by the Army Signal Corps Aviation School for WWI, this was a supply and distribution facility with various schools during World War II.
Notes	The base's origins begin with the establishment of Wilbur Wright Field on 22 May and McCook Field in November of 1917, both established by the Army Air Service as World War I installations. McCook was used as a testing field and for aviation experiments. Wright was used as a flying field (renamed Patterson Field in 1931.) [26] Wright-Patterson

Air Force Base was established in 1948 as a merger of Patterson and Wright Fields. Wishing to recognize the contributions of the Patterson family (owners of National Cash Register) the area of Wright Field east of Huffman Dam (including Wilbur Wright Field, Fairfield Air Depot, and the Huffman Prairie was renamed Patterson Field on July 6, 1931, in honor of Lt. Frank Stuart Patterson, who was killed in 1918 during a flight test of a new mechanism for synchronizing machine gun and propeller, when a rod broke during a dive from 15,000 feet causing the wings to separate from the aircraft.

Commands Air Corps Maintenance Command, 29 April 1941
Air Service Command, 17 October 1941
Tenth Air Force, 12 Feb 1942 - 17 Jun 1942
AAF Technical Service Command, 31 August 1944
(redesignated Air Technical Service Command, 1 July 1945)
Air Materiel Command, 9 March 1946

Major Units 10th Troop Carrier Group, 17 Jan 1941 - 25 May 1942
62nd Troop Carrier Group, 17 Feb 1941 - 9 Sep 1941
316th Troop Carrier Group, 1 Feb 1942 - 8 Mar 1942
478th Base Headquarters & Air Base Squadron, March 1944
(for complete listing see appendix)

Footnotes (26) *http://en.wikipedia.org/wiki/Wright-Patterson_Air_Force_Base*

Wright Field

Location Fairborn, Ohio
39 49 41N 84 02 44W
Established During World War II
Inactivated Not known
Size of Facility Not known
Mission This was the primary research and development facility for the Army Air Force.
Notes Refer to Patterson Field for much of the same history shared by the two adjacent fields.
Commands Army Air Force Materiel Center, 16 March 1942
AAF Technical Service Command, 31 August 1944
(redesignated Air Technical Service Command, 1 July 1945)
Air Materiel Command, 9 March 1946
Major Units AAF Engineer School
AAF Materiel Center
AAF Materiel Command, Headquarters
AAF Materiel Command, Government Furnished Equipment Depot
704th AAF Base Unit (Headquarters, Materiel Command - Officers)
700th AAF Base Unit (Headquarters, Materiel Command - Enlisted Personnel)
Auxiliary Fields Clinton County Army Air Field
Dayton Army Air Field

Combined: Wright Field and Patterson Fields

Both fields experienced dramatic expansion during World War II, in terms of real estate as well as structures. Employment at the fields jumped from approximately 3,700 in December 1939 to over 50,000 at the war's peak. Wright Field grew from a modest installation with approximately 30 buildings to a 2,064-acre facility with some 300 buildings and the Air Corps' first modern paved runways. The original part of the field became saturated with office and laboratory buildings and test facilities. The Hilltop area was acquired from private landowners in 1943-1944 to provide housing and services for the thousands of troops assigned to Wright Field.

The outbreak of World War II provided a crucial test for the Materiel Division. Since 1926 the Division had managed its experimental, engineering, and procurement functions with limited peacetime appropriations. Changes were required within the Division to accommodate the massive wartime Air Corps expansion program. The functions of the Division were ultimately broken into two separate commands, the Materiel Command and the Air Service Command. The Materiel Command, headquartered at Wright Field, was responsible for the procurement of airplanes and equipment in production quantities and for sustaining an accelerated program of testing and development. The Air Service Command, located on Patterson Field, assumed responsibility for all logistics functions, including maintenance and supply.

The separation of these functions soon proved cumbersome and confusing. The Army Air Forces addressed this program in August 1944 when it inactivated the two commands and reunited their functions in the newly established Air Technical Service Command. This action made Wright Field subordinate to the new headquarters at Patterson Field and had a psychologically divisive effect on the installation. To solve the problem, the portion of Patterson field from Huffman Dam through the Brick Quarters (including the command headquarters in Building 10262) was reassigned from Pattrson Field to Wright Field. To avoid confusing the two areas of Wright Field, the former Patterson Field portion was designated "Area A" of Wright Field and the original Wright Field became "Area B."

Patterson Field likewise saw the growth of hundreds of barracks and their suspporting mess halls, chapels, hospital facilities, clubs, and recreational facilities. Two densely populated housing and service areas across Highway 444, Wood City and Skyway Park, were geographically separated from the central core of Patterson Field and developed almost self-sufficient community status. (Wood City was acquired in 1924 as part of the original donation of land to the government but was used primarily as just a radio range until World War II. Skyway Park was acquired in 1943.) They supported the vast numbers of recruits who enlisted and were trained at the two fields as well as thousands of civilian laborers, especially single women recruited to work at the depot. Skyway Park was demolished after the war. Wood City was eventually transformed into Kittyhawk Center, the base's modern commercial and recreation center. [27]

Patterson field and Wright Field remained separate installations throughout World War II. As the war drew to a close, base leaders quickly recognized the need to make the most efficient use of their facilities. In 1945 they integrated the master plans for both fields and increasingly administered the functions and services of the two fields as a single installation. This practice was formalized in December 1945 with the establishment of the Army Air Forces Technical Base, Dayton, Ohio, which provided base operational support to the combined bases. On January 13, 1948, the newly created U.S. Air Force officially merged Wright and Patterson Fields to create Wright-Patterson Air Force Base. To facilitate daily management, Patterson Field became "Area C" and Skyway Park became "Area D" of the installation. Area D was donated to the State of Ohio in 1963 for the creation of Wright State University. [28]

Footnotes   (27) Ibid.

(28) *http://www.wpafb.af.mil/library/factsheets/factsheet.asp?id=6234*

Wright Field prior to WWII

Army Service Forces

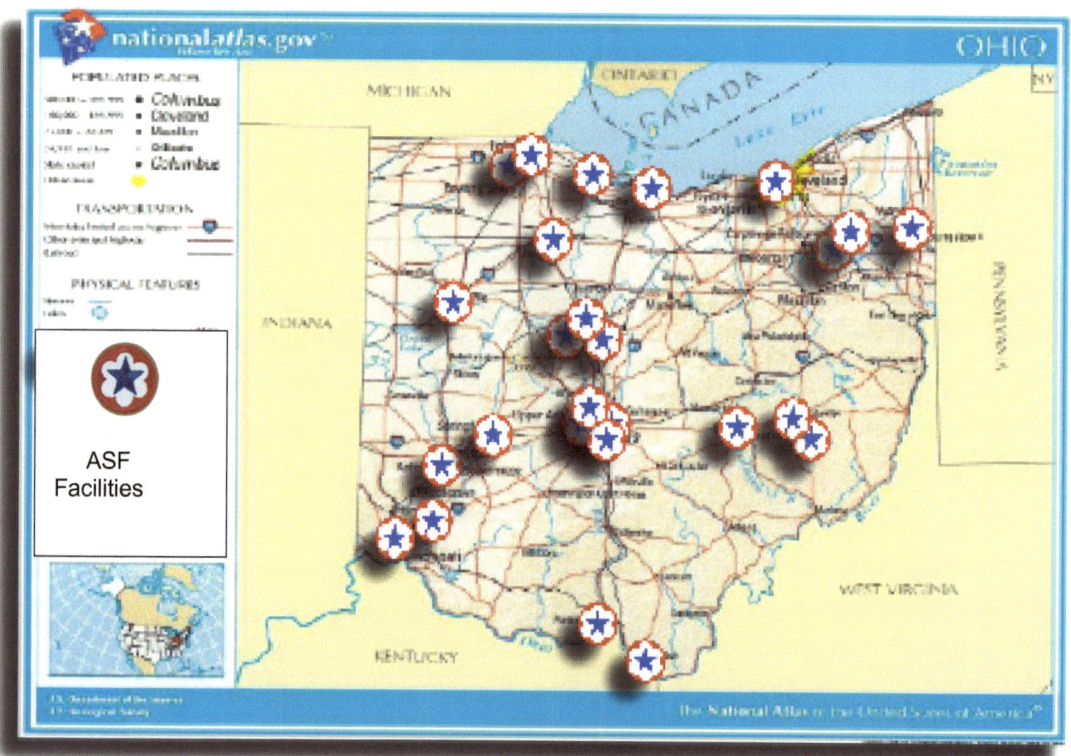

General Hospitals

Crile General Hospital

Location Parma, Ohio (NW quad: York Road / W. Pleasant Valley Road)
41 22 02N 81 45 54W
Established April 1944
Inactivated 1964
Size of Facility 153 Acres; 1,725-Bed installation
Role The site was built as a temporary facility in 1944 but following the war the facility saw several uses before ultimately becoming the home of the Cuyahoga Community College. The street pattern from the general hospital can still be witnessed relative to the community college.

Notes Another of the uses of the facility was initiated in December 1944 when a detachment of German POWs arrived from Camp Perry. They remained until the end of 1945. By the end of 1946 the role had changed again when, as with so many of the general hospitals, it became a veterans hospital (VA). This use continued until its closure in 1964 but after only 2 years it was reinvigorated as the community college.

Aerial of Crile General Hospital (looking north to south);

Placque commemorating the hospital:

- 153 Acres
- 1,725 Beds
- 15,000 soldiers cared for

Fletcher General Hospital

Location	Cambridge, Ohio	*(a little more than 3 miles north of downtown Cambridge)*
	40 04 15N 81 35 13W	
Established	14 May 1943	
Inactivated	31 March 1946	
Size of Facility	200 Acres	
Role	As with most of the general hospitals, Fletcher General Hospital had more than one life. Following its closure as a general hospital for the Army, the governor of Ohio dedicated the facility as the Cambridge State Hospital. It changed again, at least in name if not in function, to be the Cambridge Mental Health and Developmental Center.	
Notes	There were more than 120 buildings on the campus of the hospital which had its own power plant, water supply and sewer systems, as well as its own police and fire departments.	

By being in a more pastoral setting (rather than an urban setting), Fletcher General Hospital provided a psychologically calming effect for patients.

General ASF Facilities

Camp Millard

Location	Bucyrus, Ohio	(roughly a mile to the southeast of downtown Bucyrus)
Established	1 April 1942	
Inactivated	21 May 1945 - deactivated; 1 October 1945 - declared surplus	
Size of Facility	249 Acres	
Capacity:	82 Officers / 1,384 Enlisted Personnel	

Role   The services performed at Camp Millard had been provided under the simple moniker of ASF facilities at Bucyrus before the formalization of the process and the establishment of Camp Millard. The role of the camp was as a Railroad Battalion Training Center for the Transportation Corps.

Notes   The railroad shops had been established at the site of the county fairgrounds and removed after the war. The Railway Shop Battalions included the 753rd, the 755th, the 756th, 763rd, 764th, 765th, 766th and more. The Railway Workshop Battalions included the 126th, 127th, 128th, 129th, and 130th. The Hospital Train Maintenance Sections included the 117th, 118th, 119th, 120th, 121st, 138th, 139th, 140th, 141st, and 142nd. Also included was the Headquarters Detachment of the 709th Railway Grand Division.

The 758th Railway Shop Battalion was formed 6 April 1943 at Camp Harrahan (New Orleans) and inactivated 26 November 1945 at Camp Kilmer. They trained at Camp Bucyrus.

(Above) A plaque commemorating the role played by Camp Millard during the war. (Right) is the entrance to the shops area.

An aerial of the facilities at Camp Millard

Depots

Army Service Force Locations		
Depot Type	**Installation**	**Location**
General Depot	Columbus Adjutant General Depot (I)	Columbus
	Columbus Army Service Forces Depot (IV)	Columbus
Engineer Depot	Cambridge Engineer Sub-Depot (IV)	Cambridge
	Marion Engineer Depot (IV)	Marion
	Sharonville Engineer Depot (IV)	Sharonville
	Springfield Engineer Sub-Depot (IV)	Springfield
Ordnance Depot	Lordstown Ordnance Depot (IV)	Lordstown
Location	**Branch**	**Type**
Akron	Schools	Civilian Schools / University of Akron
	ASF Schools	ROTC Schools / University of Akron
	Districts	Armed Forces Induction Districts
Ashtabula	Districts	Armed Forces Induction Districts

Plants & Works

| Army Service Force Locations |||
Type Facility	Installation	Location
Ordnance Plants / Works	Buckeye Ordnance Works (IV)	Ironton
	Kings Mills Ordnance Plant (IV)	Kings Mills
	Plum Brook Ordnance Works (IV)	Sandusky
	Ravenna Ordnance Center (IV)	Ravenna
	Ravenna Ordnance Plant (IV)	Apco
	Scioto Ordnance Plant (IV)	Marion
Chemical Warfare Service Plant	Columbus CWS Plant (IV)	Columbus
	Firelands CWS Plant (IV)	Marion
	Fostoria CWS Plant (IV)	Fostoria
	Zanesville CWS Plant (IV)	Zanesville

Other ASF Installations

| Army Service Force Locations |||
Proving Ground	Installation	Location
Ordnance Proving Ground	Erie Proving Ground (Ordnance) (IV)	LaCarne

(continued overeleaf)

Communities Serving the Military

	Army Service Force Locations		
Athens	Schools	Civilian Schools / Ohio University	
	ASF Schools	ROTC Schools / Ohio University	
	Districts	Armed Forces Induction Districts	
Bucyrus	Army Service Forces	Camp Millard	
Cambridge	Army Service Forces	General Hospitals / Fletcher General Hospital	
	Depots	Engineer Depots	
Canton	Districts	Armed Forcs Induction Districts	
Chillicothe	Districts	Armed Forces Induction Districts	
	Army Service Forces	Veterans Administration Facility	
Cincinnati	Schools	Civilian Schools / University of Cincinnati	
	ASF Schools	ROTC Schools / University of Cincinnati ROTC Schools / UWilberforce University (Colored) ROTC Schools / Xavier University	
	Finance Department	Finance Offices - U.S. Army	
	Depots	Miscellaneous Activities	
	District	District Engineers	
	Districts	Armed Forces Induction Districts	
Cleveland	Schools	Civilian Schools / Western Reserve University	
	ASF Schools	ROTC Schools / Western Reserve University	
	Army Service Forces	General Hospitals / Crile General Hospital	
	District	Ordnance Procurement District	
	Districts	Armed Forces Induction Districts	
Columbus	Schools	Civilian Schools / Ohio State University	
	ASF Schools	ROTC Schools / Ohio State University	
	Army Service Forces	Army Service Forces / Columbus ASF Depot Army Service Forces / Columbus Service Cmd Shops	
	Depots / Adjutant General Depots	Columbus Adjutant General Depot Columbus ASF Depot Columbus Chemical Warfare Plant Habus Chemical Warfare Plant	
	Districts	Armed Forces Induction Districts	
	Finance Department	Columbus Finance Offices, U.S. Army	
	Real Estate Districts	Real Estate Division	
	Quartermaster	Quartermaster Market Centers	
	District	Officer Procurement Districts	
	Divisions	Engineer Divisions	
	Pools	Armed Forces Induction Districts	

Army Service Force Locations

Location	Branch	Type
[Fort Hayes]	Army Service Forces	Fort Hayes Reception Center
	Inspector General	Officer Procurement Pools
	Pools	Judge Advocate, Officer Replacement Pools
Dayton	Schools	Signal Corps Schools
		Civilian Schools / University of Dayton
	ASF Schools	ROTC Schools / University of Dayton
	Finance Department	Finance Offices, U.S. Army
	Depots	MIscellaneous Activities
	Pools	Signal Corps Officer Replacement Pool
	District	District Engineers
	Districts	Armed Forces Induction Districts
[Wright Field]	Army Service Forces	Laboratories (Photographic)
	Depots	Miscellaneous Activities
	Pools	Signal Corps Officer Replacement Pool
East Liverpool	Districts	Armed Forces Induction Districts
Fostoria	Depots	Chemical Warfare Plants
Gambier	Schools	Civilian Schools / Kenyon College
Hamilton	Districts	Armed Forces Induction Districts
Ironton	Depots	Plants & Works / Buckeye Ordnance Works
Kings Mills	Depots	Plants & Works / Kings Mills Ordnance Plant
La Carne	Army Service Forces	Proving Grounds
Lima	Districts	Armed Forces Induction Districts
Mansfield	Districts	Armed Forces Induction Districts
Marietta	Districts	Armed Forces Induction Districts
Marion	Depots	Engineer Depots / Marion Engineer Depot
		Plants & Works / Scioto Ordnance Plant
	Districts	Armed Forces Induction Districts
New Concord	Schools	Civilian Schools / Muskingum College
Port Clinton	Army Service Forces	Prisoner of War Camp
[Erie Proving Ground]	Ordnance Depots	Erie Ordnance Depot
Portage	Depots	Ordnance Depots / Portage Ordnance Depot
Portsmouth	Districts	Armed Forces Inductino Districts
Ravenna	Army Service Forces	Army Technical Service Center
	Depots	Plants & Works / Ravenna Ordnance Plant
Sandusky	Depots	Plants & Works / Plum Rock Ordnance Works
	Districts	Armed Forces Induction Districts
Sharonville	Depots	Engineer Depots
Springfield	Depots	Engineer Depots
	Districts	Armed Forces Induction Districts
Steubenville	Districts	Armed Forces Induction Districts
Tiffin	Schools	Civilian Schools / Heidelberg College

U.S. Army Installations – World War II: Fifth Service Command

Location	Branch	Type
Toledo	Depots	Medical Depots / Toledo Medical Supply Depot
		Ordnance Depots / Rossford Ordnance Depot
	Districts	Armed Forces Induction Districts
Warren	Depots	Ordnance Depots / Lordstown Ordnance Depot
Wilberforce	Schools	Civilian Schools / Wilberforce University (Colored)
Youngstown	Districts	Armed Forces Induction Districts
Zanesville	Depots	Chemical Warfare Plants / Zanesville CWS Plant
	Districts	Armed Forces Induction Districts

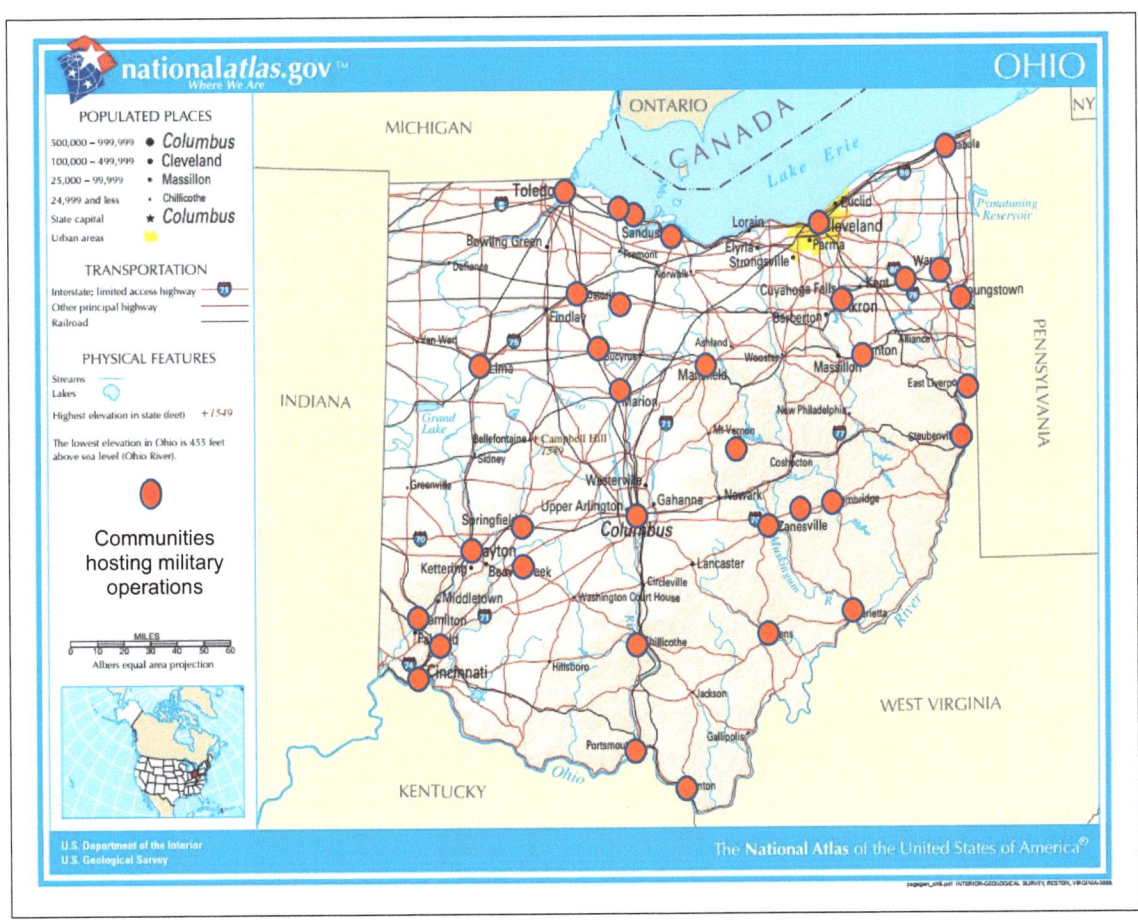

Prisoner of War Camps

Camp Perry Base POW Camp
Fort Knox, Kentucky

Type Facility	Base POW Camp
Period	Italians were held at Camp Perry from October 1943 through June 1944 and beginning in July 1944 through February 1946, the German POWs replaced the Italians
Location	Located at the Camp Perry post
Size / Capacity	Capacity varied: 1,000 / 2,945 / 2,950 / 4,355 / 3,145
Num. Prisoners-Avg / Max	4,020 German POW average over 600 days / 5,795 peak in Sept 1945
	1,341 Italian POW average over 240 days / 2,013 peak in March 1944
Nationality	Italian (first); replaced by Germans
MP / Service Units	1591st Service Command Unit: Prisoner of War Camp

Branches

Bowling Green Branch Camp, Bowling Green, OH
German POWs from September 1944 through November 1945.
455 average over 180 days / 581 peak in October 1945

Celina Branch Camp, Celina, OH
German POWs between September 1944 through October 1945.
260 average over 60 days / 428 peak in October 1945

Crile General Hospital Branch Camp, Cleveland, OH
Crile held German POWs from January 1945 and February 1946.
250 average over 390 days / 370 peak in September 1945

Defiance Branch Camp, Defiance, OH
German POWs from September 1944 through November 1945.
300 average over 450 days / 659 peak in October 1945

Fletcher General Hospital Branch Camp, Cambridge, OH
German POWs from February 1945 through February 1946.
181 average over 360 days / 232 peak in September 1945

Fort Hayes Branch Camp, Columbus, OH
German POWs from February 1945 and January 1946.
337 average over 360 days / 450 peak in August 1945

Marion Engineer Depot Branch Camp, Marion, OH
German POWs from January 1945 through February 1946.
289 average over 420 days / 507 peak in September 1945

Wilmington Branch Camp, Wilmington, OH
German POWs from September 1944 through November 1945.
141 average over 90 days / 222 peak in August 1945

Wright Field Branch Camp, Dayton, OH
German POWs from September 1944 through November 1945.

666 average over 90 days / 783 peak in February 1946

USFR Chillicothe POW Facility
Chillicothe, OH
German POWs from September 1944 through November 1945.

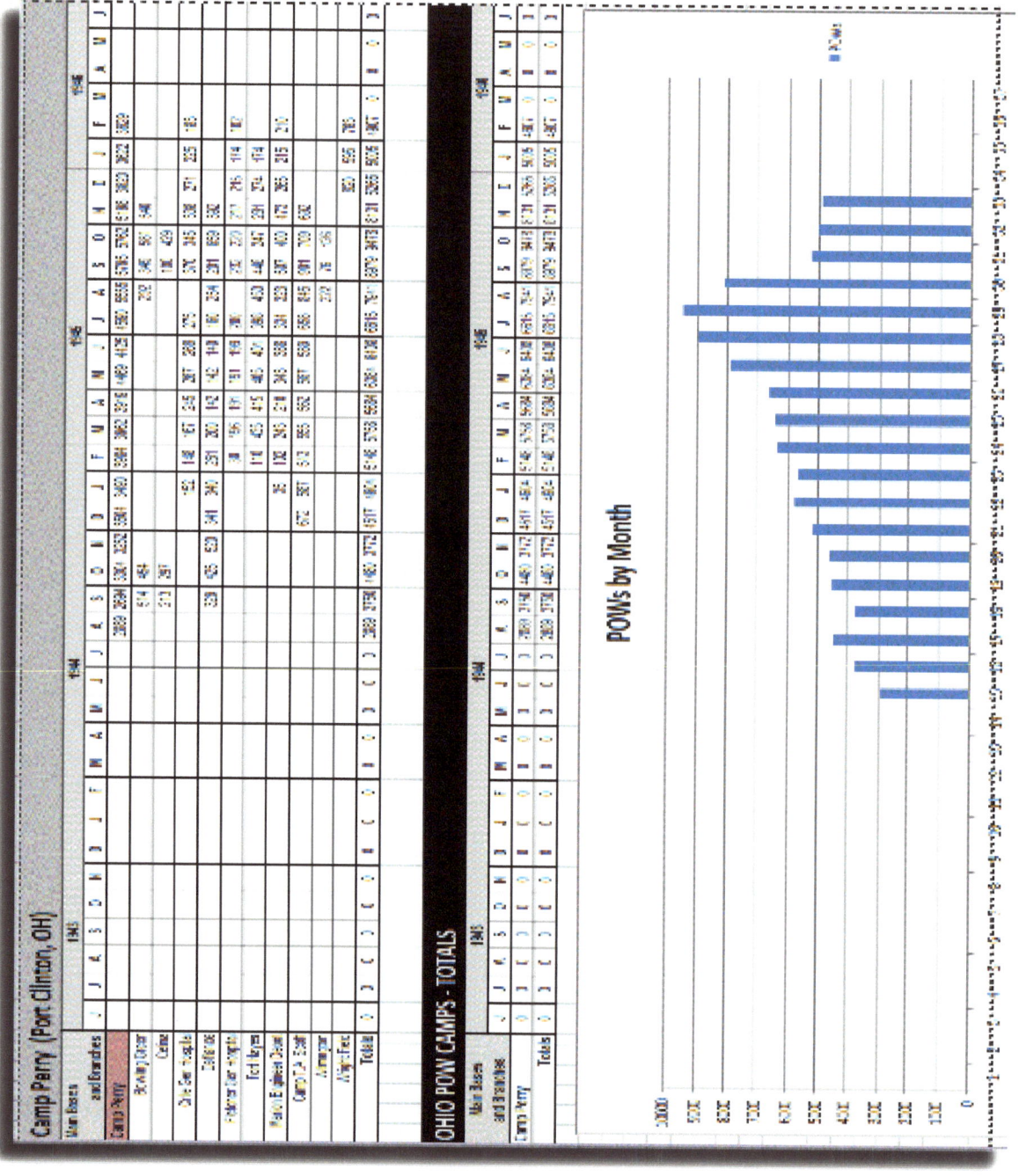

West Virginia

West Virginia

Army Ground Force

Camp Dawson

Location	Kingwood, West Virginia *(near Kingwood, in Preston County, Dunkard Bottom)*
Established	7 May 1909
Inactivated	Remains an active army post, not federalized, but operating as a National Guard training center.
Size of Facility	249 Acres (initially); additional acquisitions have brought the total to 4,177 Acres
Role	Camp Dawson had been a unit training facility since the activation of the camp until the start of World War I. The camp was not used again until 1928 when it was re-established as a training site for the West Virginia State Militia. From that time, troops trained at Camp Dawson regularly but with the advent of WWII, the government leased the camp for use as a POW camp.

Army Air Force

There were only a few army air fields in the state of West Virginia during World War II and none of those played a major role in the war effort. They were not tactical installations nor were they training fields, per se. There is an airfield aligned with Camp Dawson but that did not exist during World War II, in fact, it wasn't built until 1970 and was replaced in 1975.

There were army airfields at Lewisburg, Cumberland, Elkins, Moundsville and Buckhannon but none of these with the exception of Greenbrier Army Air Field at Lewisburg had a documented role during the war. The others were "used by the Army." One source simply states that Greenbriar was "used by the Army Air Force" whereas another indicates that the field was used by the First Air Force - a somewhat more defined role, but still without meaningful detail.

Cumberland Field

Location Cumberland, West Virginia
38 53 21N   79 51 25W
No details found regarding the former army air Field

Greenbriar Army Air Field

Location Lewisburg, West Virginia
37 51 30N   80 23 56W
The pre-existing public airport (opened 1 September 1930) was contracted for in 1942 at the same time as the federalization of the Greenbrier Hotel to be the Ashford General Hospital, and subsequently a POW Camp. It was returned to civilian use in 1946.

It is believed to have been 165 acres in 1945 and operated under the First Air Force, but the facility was expanded subsequently to 472 acres and operated as the Greenbrier Valley Airport. Today the area is a very large golf course with no signs of an air field having been there.

Harpers Field

Location Elkins, West Virginia
38 57 22N   79 47 19W
No details found regarding the former army air Field

Langin Field

Location Moundsville, West Virginia
No details found regarding the former army air Field

Lewis Field

Location Buckhannon, West Virginia
No details found regarding the former army air Fi

Army Service Forces

General Hospitals

Ashford General Hospital

Location	White Sulphur Springs, West Virginia
Established	August 1942
Inactivated	30 June 1946 - closed
Size of Facility	7,000 Acres; 1,725-Bed installation
Role	What became the Ashford General Hospital had been the Greenbrier Hotel in White Sulphur Springs, West Virginia, prior to the acquisition by the Army. This was a general hospital operated by the Army employing 45 doctors, 100 nurses, and 500 enlisted men. There were also 200 WACs, 35 civilian nurses and more than 500 civilian employees in addition to the volunteers available through the Red Cross.
Notes	The facility was purchased from the Chesapeake and Ohio Railroad for $3.3 million and renamed as one of the Army's general hospitals. Over the four year period of service, more than 24,000 patients were treated at the facility. At the end of the Army's use of the hotel / hospital, it was sold back to the Chesapeake & Ohio Railroad for the same price paid by the Army. [28] [29]

Footnotes
(28) Keefer, Louis E., at http://www.wvencyclopedia.org/articles/304
(29) Fallon, Paul, "When the Greenbrier Was A Hospital," at http:www.charlestondailymail.com/News/statenews/201012091090

Newton D. Baker General Hospital

Location	Martinsburg, West Virginia
Established	28 January 1944
Inactivated	20 June 1946 (last patient discharged)
Size	1,806 Beds
Role	Newton D. Baker General Hospital was one of the Army's Zone of the Interior general hospitals.

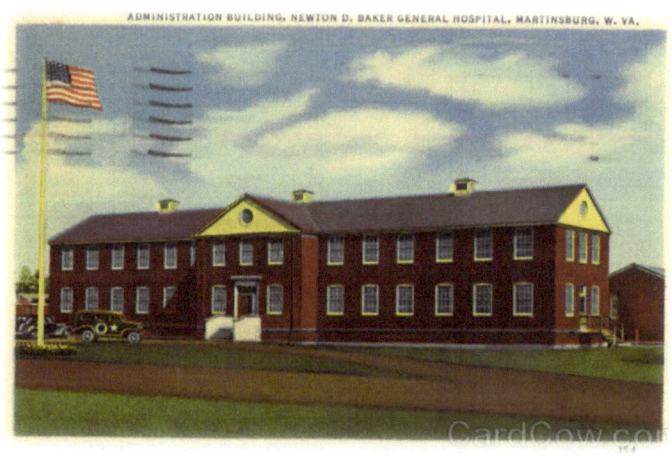

Plants & Works

Kanawha Chemical Warfare Service Plant (IV)

Location	South Charleston, West Virginia
Established	1 April 1942
Inactivated	21 May 1945 - deactivated; 1 October 1945 - declared surplus
Role	The services performed at Camp Millard had been provided under the simple moniker of ASF facility.

Typical ordnance igloo (sofrage)

Marshall Chemical Warfare Service Plant (IV)

Location   New Martinsville, West Virginia
Established
Inactivated
Size of Facility
Role   One of the government-owned, privately operated CWS plants run by E.I. du Pont Nemours & Co., Inc. The plant constructing / outfitting cost over $5 million.

Morgantown Ordnance Works (IV)

Location   Morgantown, West Virginia
Established   Not known
Inactivated   Not known
Size of Facility Not known
Role   One of the government-owned, privately operated CWS plants.

West Virginia Ordnance Works (IV)

Location   Point Pleasant, West Virginia   (Headquarters at Camp Conley)
Established   October 1942
Inactivated   1945 - deactivated; 27 February 1946 - Corps of Engineers announced pending sale
Size of Facility 8,323 Acres - expanded to 9,000 Acres
Role   The plant produced 720,000 pounds of TNT every 24 hours and then shipped it to ordnance plans that utilized the TNT into the various ordnance types. The land use pattern that shows the containment igloos for the ordnance is still visible through Google Earth.

Army Service Force Locations

Location	Branch	Type
Beckley	Districts	Armed Forces Induction Districts
Bluefield	Districts	Armed Forces Induction Districts
Charleston	Districts	Armed Forces Induction Districts
	Depots	Chemical Warfare Plants / Kanawha CWS Plant
Clarksburg	Districts	Armed Forces Induction Districts
Huntington	Districts	Armed Forces Induction Districts
	District	District Engineers
Institute	Schools	Civilian Schools / West Virginia State College (Colored)
	ASF Schools	ROTC Schools / West Virginia State College (Colored)
Lewisburg	School	ROTC Schools / Greenbriar Military Academy
Logan	Districts	Armed Forces Induction Districts
Martinsburg	Districts	Armed Forces Induction Districts
	Army Service Forces	General Hospitals / Newton D. Baker General Hospital
Morgantown	Districts	Armed Forces Induction Districts
	Schools	Civilian Schools / West Virginia University
	ASF Schools	ROTC Schools / West Virginia University
	Depots	Plants & Works / Morgantown Ordnance Works
New Martinsville	Depots	CWS Plants / Marshall CWS Plant
Parkersburg	Districts	Armed Forces Induction Districts
Point Pleasant	Depots	Plants & Works / West Virginia Ordnance Works
Welch	Districts	Armed Forces Induction Districts
Wheeling	Districts	Armed Forces Induction Districts
White Sulphur Springs	Army Service Forces	General Hospitals / Ashford General Hospital
	Depots	Field Printing Plants
	Schools	Medical Schools
	Army Service Forces	Prisoner of War Camp / Ashford General Hospital

Prisoner of War Camps POW

Camp Ashford Base POW Camp
White Sulphur Springs, West Virginia

Type Facility	Base POW Camp
Location	Located at a former CCC camp
Size / Capacity	Capacity varied: 1,000 / 1,020 / 1,000; 165 acres at former CCC camp
Num. Prisoners-Avg / Max	573 German POW average over 90 days / 680 peak in September 1943
	891 Italian POW average over 1,020 days / 1,561 peak in February 1946
Nationality	Italian (first); replaced by Germans
MP / Service Units	486th Military Police Escort Guard Company (ASF)
	1514th Service Command Unit: Prisoner of War Camp
Branches	***Ashford General Hospital Branch Camp***, Greenbriar Hotel (see below).
	Camp Dawson Branch Camp, Kingwood, West Virginia
	175 Italian prisoners were held at the WV National Guard site

Ashford General Hospital Base POW Camp / PMI
White Sulphur Springs, West Virginia

Type Facility	At one point a Base Camp / a Banch Camp / a Penal-Medical Institution
Period	Germans were held here between September 1943 and December 1943
Location	These prisoners were assigned to the hospital
Size / Capacity	
Num. Prisoners-Avg / Max	46 German POW average over 120 days / 47 peak in September 1943
Nationality	Germans
MP / Service Units	Not know

Newton D. Baker General Hospital POW Camp / PMI
Martinsburg, West Virginia

Type Facility	Penal / Medical Institution
Period	Not known
Location	Prisoners held at the hospital (where they performed maintenance work)
Size / Capacity	Capacity varied: 1,000 / 2,945 / 2,950 / 4,355 / 3,145
Num. Prisoners-Avg / Max	242 German POW average over 540 days / 363 peak in Sept 1945
Nationality	Germans
MP / Service Units	Not known

Logan General Hospital POW Camp / PMI
Logan, West Virginia

Type Facility	Penal / Medical Institution (POW Camp)
Period	Not Known
Location	Located at Logan General Hospital
Size / Capacity	Not Known
Num. Prisoners-Avg / Max	1 German POW average over 30 days / 1 peak in March 1945
Nationality	Germans
MP / Service Units	Not Known

Rainelle Branch POW Camp [30]
Rainelle, West Virginia

Type Facility	Branch POW Camp
Period	Not Known
Location	Located at Logan General Hospital
Size / Capacity	Not Known
Num. Prisoners-Avg / Max	49 German POW average over 120 days / 47 peak in September 1943
Nationality	German
Footnotes	(30) *In contact with several libraries in Charleston and Rainelle, the librarians (usually a great source for information, could neither recall such a facility nor could they find a reference at their libraries. However, the Station List of the Army of the United States so cites such a facility; therefore, despite scant information it is included*

Appendices

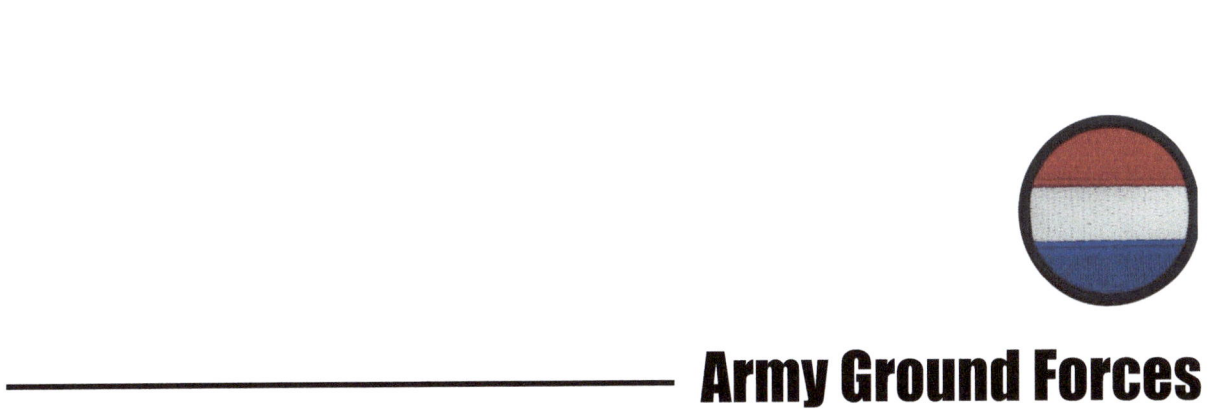

Army Ground Forces

Fifth Service Command
Columbus Ohio

Army Ground Force Posts

Camp Atterbury (II)
Edinburgh, Indiana

Army Service Forces

Wakeman General & Convalescent Hospital (see Wakeman Hospital)
Wakeman General Hospital
Wakeman Hospital Center
War Department Personnel Center (see War Department...)
Post Photographic Laboratory (ASF)
Sub-School for Bakers & Cooks
Tire Collection Center (I Act)
Fifth Service Command Signal Corps Photographic Sub-Laboratory (SOS)
8th General Hospital (SOS)
18th Hospital Center (SOS)
18th Hospital Center Band (SOS)
27th Medical Depot Company (ASF)
28th Medical Depot Company (ASF)
36th Portable Surgical Hospital (ASF)
49th Field Hospital (ASF)
50th Field Hospital (ASF)
62nd General Hospital (ASF)
72nd General Hospital
73rd General Hospital
118th Station Hospital (500-Bed) (ASF)
132nd Station Hospital (100-Bed)
195th Station Hospital (SOS)
237th Station Hospital (250-Bed) (ASF)
264th Military Police Company (ZI) (SOS)
317th Station Hospital (750-Bed) (ASF)
322nd Military Escort Guard Detachment
334th Military Escort Guard Detachment
337th Military Escort Guard Detachment
350th Military Escort Guard Detachment
451st ASF Band
669th Quartermaster Service Battalion, HQ & HQ Detachment (Colored)
979th Quartermaster Service Company
980th Quartermaster Service Company
981st Quartermaster Service Company
982nd Quartermaster Service Company
729th Military Police Battalion
749th Sanitation Company (Medium) (Colored) (ASF)

- 750th Sanitation Company (Medium) (Colored) (ASF)
- 751st Sanitation Company (Medium) (Colored) (ASF)
- 797th Sanitation Company (Medium) (ASF)
- 798th Military Police Battalion
- 1506th Service Command Unit: Sub-School for Bakers & Cooks #3 (SOS)
- 1512th Service Command Unit: Fifth Service Command Motor Pool
- 1513th Service Command Unit: Ordnance Service Command Shop A, District #2 (SOS)
- 1513th Service Command Unit: Ordnance Service Command Shop (ASF)
- 1537th Service Command Unit: Camp Atterbury Internment Camp (SOS)
- 1560th Service Command Unit: Station Complement (ASF) (includes WAC personnel)
- Fifth Service Command Motor Pool
- Medical Department Enlisted Technicians School (WAC)
- Packing Squad
- Prisoner of War Camp
- Special Training Unit
- Wakeman General & Convalescent Hospital (includes WAC personnel)
- Wakeman Hospital Center
- War Department Personnel Center
- Basic Training Section
- Fifth Service Command Vehicle Distribution Park
- 1560th Service Command Unit: Station Complement, WAC HQ Detachment
- 1560th Service Command Unit: Station Complement, WAC Medical Det (Colored)
- 1562nd Service Command Unit: Sub-School for Bakers & Cooks (ASF)
- 1584th Service Command Unit: Special Training Unit
- 1596th Service Command Unit: Fifth Service Command Packing Squad
- 3555th Service Command Unit: Casual Company (ASF)
- 3561st Service Command Unit: WAC Station Complement (ASF)

Army Ground Forces

- 8th HQ & HQ Detachment, Special Troops, Second Army (AGF)
- 8th HQ & HQ Detachment, Special Troops, Second Army (Type A) (AGF)
- 9th Tank Destroyer Group, HQ & HQ Company (AGF)
- 608th Tank Destroyer Battalion
- 610th Tank Destroyer Battalion (Towed)
- 11th Ordnance Battalion, HQ & HQ Detachment (AGF)
- 895th Ordnance Heavy Auto Maintenance Company
- 3532nd Ordnance Medium Auto Maintenance Company
- 16th Field Artillery Observation Battalion, Battery A (AGF)
- 30th Infantry Division, Headquarters (AGF)
- 30th Infantry Division, Special Troops, Headquarters
- Headquarters Company, 30th Infantry Division watchd Medical & Ct
- 30th Infantry Division, Military Police Platoon
- 30th Infantry Division Band
- 730th Ordnance Light Maintenance Company
- 30th Quartermaster Company
- 30th Signal Company
- 117th Infantry Regiment
- 119th Infantry Regiment

- 120th Infantry Regiment
- 30th Division Artillery, HQ & HQ Battery (Motorized)
- 113th Field Artillery Battalion (Motorized) (155mm Howitzer, Truck-D)
- 118th Field Artillery Battalion (Motorized) (105mm Howitzer, Truck-D)
- 197th Field Artillery Battalion (Motorized) (105mm Howitzer, Truck-D)
- 230th Field Artillery Battalion (Motorized) (105mm Howitzer, Truck-D)
- 30th Reconnaissance Troop, Mechanized
- 105th Engineer Combat Battalion
- 106th Medical Battalion
- 30th Signal Construction Battalion (AGF)
- 31st Signal Construction Battalion (AGF)
- 35th Evacuation Hospital, (Mtz) (AGF)
- 35th Evacuation Hospital, Semi Mobile (AGF)
- 39th Evacuation Hospital, (Mtz) (AGF)
- 39th Evacuation Hospital, Semi Mobile (AGF)
- 42nd Signal Construction Battalion (Colored) (AGF)
- 44th Evacuation Hospital (Mtz) (AGF)
- 74th AGF Band (Colored) (AGF)
- 75th Ordnance Medium Maintenance Battalion, HQ & HQ Detachment (Q) (AGF)
- 75th Quartermaster Company (AGF)
- 82nd AGF Band (AGF)
- 83rd Division Artillery Band
- 83rd Infantry Division, Headquarters (AGF)
- 83rd Infantry Division, Headquarters Company
- 83rd Infantry Division, Military Police Platoon
- 83rd Cavalry Reconnaissance Troop
- 83rd Signal Company
- 329th Infantry Regiment (less Band)
- 330th Infantry Regiment
- 331st Infantry Regiment (less Band)
- 83rd Division Artillery, Headquarters
- 83rd Division Artillery, Headquarters Battery
- 322nd Field Artillery Battalion (105mm Howitzer, Truck-D)
- 323rd Field Artillery Battalion (105mm Howitzer, Truck-D)
- 324th Field Artillery Battalion (155mm Howitzer, Truck-D)
- 908th Field Artillery Battalion (105mm Howitzer, Truck-D)
- 308th Engineer Battalion
- 308th Medical Battalion
- 83rd Quartermaster Company
- 783rd Ordnance Light Maintenance Company
- 100th Quartermaster Bakery Battalion, HQ & HQ Company, Medical Det & Co C (Colored)
- 106th Infantry Division, Headquarters
- HQ Company, 106th Infantry Division, Watchd Med & Chap
- 106th Infantry Division, Military Police Platoon
- 106th Infantry Division Band
- 806th Ordnance Light Maintenance Company

- 106th Quartermaster Company
- 106th Signal Company
- 422nd Infantry Regiment
- 423rd Infantry Regiment
- 424th Infantry Regiment
- 106th Division Artillery, HQ & HQ Battery (Mtz)
- 589th Field Artillery Battalion (Mtz) (105mm Howitzer, Truck-D)
- 590th Field Artillery Battalion (Mtz) (105mm Howitzer, Truck-D)
- 591st Field Artillery Battalion (Mtz) (105mm Howitzer, Truck-D)
- 592nd Field Artillery Battalion (Mtz) (155mm Howitzer, Truck-D)
- 106th Reconnaissance Troop, Mechanized
- 81st Engineer Combat Battalion
- 331st Medical Battalion
- 101st Infantry Battalion (Separate) (w/Medical Section & Transportation Section) (AGF)
- 122nd Ordnance Medium Maintenance Company (AGF)
- 122nd Ordnance Light Maintenance Company (AGF)
- 141st Ordnance Medium Maintenance Company (AGF)
- 164th Signal Photo Company -General Assignment Unit #2 (AGF)
- 168th Signal Photo Company - 1 Combat Assignment Unit & 1 Combat Lab Unit
- 193rd Signal Repair Company (less 4 Radio Repair Sections) (AGF)
- 196th AGF Band (Colored) (AGF)
- 206th Quartermaster Gas Supply Battalion, Company D (AGF)
- 210th Ordnance Battalion, HQ & HQ Detachment (AGF)
- 141st Ordnance Medium Maintenance Company
- 377th Ordnance Medium Maintenance Company
- 417th Ordnance Light Maintenance Company, Field A
- 963rd Ordnance Heavy Auto Maintenance Company
- 227th AGF Band (Colored) (AGF)
- 234th AGF Band
- 249th Quartermaster Service Battalion, HQ & HQ Detachment & Cos C & D (AGF)
- 301st Signal Operations Battalion (AGF)
- 306th Quartermaster Railhead Company (Colored) (AGF)
- 308th Quartermaster Railhead Company (Colored) (AGF)
- 324th AGF Band
- 330th Infantry Regiment Band
- 349th Ordnance Motor Transport Company (Q) (AGF)
- 365th Infantry Regiment Band (AGF)
- 365th Infantry Regiment (AGF)
- 366th Infantry Regiment (Colored) (AGF)
- 371st Ordnance Medium Maintenance Company (AGF)
- 379th Quartermaster Truck Company (Colored) (AGF)
- 428th Medical Ambulance Battalion (Colored) (AGF)
- 428th Medical Battalion (Separate) (Colored) (AGF)
- 587th Ambulance Company, Motor (Separate)
- 588th Ambulance Company, Motor (Separate)
- 589th Ambulance Company, Motor (Separate)
- 457th Engineer Depot Company (AGF)

574th Quartermaster Railhead Company (Colored) (AGF)
597th Field Artillery Battalion (AGF)
608th Tank Destroyer Battalion (AGF)
673rd Engineer Topographic Company (AGF)
772nd Tank Battalion (AGF)
773rd Tank Destroyer Battalion (AGF)
895th Ordnance Heavy Maintenance Company (Q) (AGF
895th Ordnance Heavy Auto Maintenance Company (AGF
3479th Ordnance Medium Maintenance Company (Q) (AGF)
3532nd Ordnance Medium Auto Maintenance Company (AGF)

Army Air Forces

1219th Military Police Company (Aviation) (AAF)
1220th Military Police Company (Aviation) (AAF)
1231st Military Police Company (Aviation) (AAF)
1232nd Military Police Company (Aviation) (AAF)
1233rd Military Police Company (Aviation) (AAF)
1234th Military Police Company (Aviation) (AAF)
1235th Military Police Company (Aviation) (AAF)
1236th Military Police Company (Aviation) (AAF)
1237th Military Police Company (Aviation) (AAF)
1238th Military Police Company (Aviation) (AAF)
1239th Military Police Company (Aviation) (AAF)
1240th Military Police Company (Aviation) (AAF)
1241st Military Police Company (Aviation) (AAF)
1242nd Military Police Company (Aviation) (AAF)

Fort Benjamin Harrison (I)
Indianapolis, Indiana

Army Service Forces

Fifth Corps Area Induction Station (Fort Benjamin Harrison)
Fort Benjamin Harrison Reception Center
Fort Benjamin Harrison School for Bakers & Cooks
Billings General Hospital
Schoen Field (see Schoen Field)
Finance Replacement Training Center Band
Finance Department Replacement Training Center (ASF)
    Finance Department Replacement Training Center, 1st Finance Training Bn
    Finance Department Replacement Training Center, 2nd Finance Training Bn
    Finance Department Replacement Training Center, 3rd Finance Training Bn
    Provisional Finance Training Battalion
Veterinary Detachment HH (ASF)
Veterinary Detachment NN (Food Inspection) (ASF)
Post Photographic Laboratory (ASF)
Finance School
Finance Training Center (ASF)
Army Service Forces Training Center (IV Act)
    Army Finance School

Organizational Chart / Gantt-style Table

Unit / Course	Presence
Finance Department Officer Replacement Pool	✗
Officer Candidate School	✗
Unit Training (Finance)	✗
Pre-Activation Training (Finance)	✗
Basic Finance Enlisted Course (WAC)	✗
Advanced Enlisted Course (Finance)	✗
Officer Candidate Course (Finance)	✗
Officers Disbursing Course (Finance)	✗
Advanced Fiscal Officers Course	✗
Civilian Pay Roll Administration Course	✗
Terminating Accounting & Auditing Course	✗
Midwestern Branch, US Disciplinary	✗
Field Printing Plant	✗
Provost Marshal Area #6	✗
Reception Center	✗
Reception Station #6	✗
5th Corps Area Lab	✗
1st Finance Section (SOS)	✗
2nd Finance Section (SOS)	✗
3rd Finance Section (SOS)	✗
4th Finance Section (SOS)	✗
5th Finance Section (SOS)	✗
6th Finance Section (SOS)	✗
7th Finance Section (SOS)	✗
8th Finance Section (SOS)	✗
9th Finance Section (SOS)	✗
10th Finance Section (SOS)	✗
11th Finance Section (SOS)	✗
12th Finance Section (SOS)	✗
13th Finance Section (SOS)	✗
14th Finance Section (SOS)	✗
15th Finance Section (SOS)	✗
17th Finance Disbursing Section (ASF)	✗
18th Finance Disbursing Section (ASF)	✗
19th Finance Disbursing Section (ASF)	✗
20th Finance Disbursing Section (ASF)	✗
41st Finance Disbursing Section (ASF)	✗
42nd Finance Disbursing Section (ASF)	✗
43rd Finance Disbursing Section (ASF)	✗
44th Finance Disbursing Section (ASF)	✗
45th Finance Disbursing Section (ASF)	✗
47th Finance Disbursing Section (ASF)	✗
48th Finance Disbursing Section (ASF)	✗
49th Finance Disbursing Section (ASF)	✗
50th Finance Disbursing Section (ASF)	✗
53rd Finance Disbursing Section (ASF)	✗
65th Finance Disbursing Section (ASF)	✗

- 66th Finance Disbursing Section (ASF)
- 67th Finance Disbursing Section (ASF)
- 68th Finance Disbursing Section (ASF)
- 69th Finance Disbursing Section (ASF)
- 70th Finance Disbursing Section (ASF)
- 71st Finance Disbursing Section (ASF)
- 72nd Finance Disbursing Section (ASF)
- 73rd Finance Disbursing Section (ASF)
- 74th Finance Disbursing Section (ASF)
- 75th Finance Disbursing Section (ASF)
- 76th Finance Disbursing Section (ASF)
- 77th Finance Disbursing Section (ASF)
- 78th Finance Disbursing Section (ASF)
- 79th Finance Disbursing Section (ASF)
- 80th Finance Disbursing Section (ASF)
- 81st Finance Disbursing Section (ASF)
- 82nd Finance Disbursing Section (ASF)
- 83rd Finance Disbursing Section (ASF)
- 84th Finance Disbursing Section (ASF)
- 85th Finance Disbursing Section (ASF)
- 86th Finance Disbursing Section (ASF)
- 87th Finance Disbursing Section (ASF)
- 85th Quartermaster Battalion, Company A (Light Maintenance)
- 89th Finance Disbursing Section (ASF)
- 90th Finance Disbursing Section (ASF)
- 90th Medical Composite Section (Food Inspection) (ASF)
- 91st Finance Disbursing Section (ASF)
- 92nd Finance Disbursing Section (ASF)
- 93rd Finance Disbursing Section (ASF)
- 94th Finance Disbursing Section (ASF)
- 94th Medical Composite Section (Food Inspection) (ASF)
- 95th Finance Disbursing Section (ASF)
- 96th Finance Disbursing Section (ASF)
- 97th Finance Disbursing Section (ASF)
- 98th Finance Disbursing Section (ASF)
- 99th Finance Disbursing Section (ASF)
- 100th Finance Disbursing Section (ASF)
- 101st Finance Disbursing Section (ASF)
- 102nd Finance Disbursing Section (ASF)
- 103rd Finance Disbursing Section (ASF)
- 104th Finance Disbursing Section (ASF)
- 105th Finance Disbursing Section (ASF)
- 106th Finance Disbursing Section (ASF)
- 107th Finance Disbursing Section (ASF)
- 108th Finance Disbursing Section (ASF)
- 109th Finance Disbursing Section (ASF)
- 110th Finance Disbursing Section (ASF)

- 111th Finance Disbursing Section (ASF)
- 112th Finance Disbursing Section (ASF)
- 113th Finance Disbursing Section (ASF)
- 114th Finance Disbursing Section (ASF)
- 115th Finance Disbursing Section (ASF)
- 116th Finance Disbursing Section (ASF)
- 117th Finance Disbursing Section (ASF)
- 118th Finance Disbursing Section (ASF)
- 119th Finance Disbursing Section (ASF)
- 120th Finance Disbursing Section (ASF)
- 121st Finance Disbursing Section (ASF)
- 122nd Finance Disbursing Section (ASF)
- 123rd Finance Disbursing Section (ASF)
- 124th Finance Disbursing Section (ASF)
- 125th Finance Disbursing Section (ASF)
- 126th Finance Disbursing Section (ASF)
- 127th Finance Disbursing Section (ASF)
- 128th Finance Disbursing Section (ASF)
- 129th Finance Disbursing Section (ASF)
- 143rd Finance Disbursing Section (ASF)
- 144th Finance Disbursing Section (ASF)
- 145th Finance Disbursing Section (ASF)
- 146th Finance Disbursing Section (ASF)
- 147th Finance Disbursing Section (ASF)
- 148th Finance Disbursing Section (ASF)
- 149th Finance Disbursing Section (ASF)
- 150th Finance Disbursing Section (ASF)
- 151st Finance Disbursing Section (ASF)
- 152nd Finance Disbursing Section (ASF)
- 153rd Finance Disbursing Section (ASF)
- 154th Finance Disbursing Section (ASF)
- 162nd Finance Disbursing Section (ASF)
- 163rd Finance Disbursing Section (ASF)
- 164th Finance Disbursing Section (ASF)
- 165th Finance Disbursing Section (ASF)
- 166th Finance Disbursing Section (ASF)
- 167th Finance Disbursing Section (ASF)
- 168th Finance Disbursing Section (ASF)
- 169th Finance Disbursing Section (ASF)
- 195th Finance Disbursing Section (ASF)
- 196th Finance Disbursing Section (ASF)
- 197th Finance Disbursing Section (ASF)
- 198th Finance Disbursing Section (ASF)
- 199th Finance Disbursing Section (ASF)
- 201st Infantry Regiment
- 229th Finance Disbursing Section
- 230th Finance Disbursing Section

(Gantt-style chart showing activity periods for various Finance Disbursing Sections: 231st, 236th, 237th, 238th, 239th, 241st, 242nd, 243rd, 244th, 245th, 246th, 247th, 248th, 249th, 250th, 251st, 252nd, 253rd, 254th, 255th, 256th, 257th, 258th, 259th, 260th, 261st, 262nd, 263rd, 264th, 265th, 266th, 267th, 268th, 269th, 270th, 271st, 272nd, 273rd, 274th, 301st, 302nd, 303rd, 304th, 305th, 306th, 307th, 308th Finance Disbursing Section.)

U.S. Army Installations - World War II: A Research Reference

- 309th Finance Disbursing Section
- 310th Finance Disbursing Section
- 311th Finance Disbursing Section
- 312th Finance Disbursing Section
- 313th Finance Disbursing Section
- 314th Finance Disbursing Section
- 342nd ASF Band
- 729th Military Police Battalion (Z of I) (ASF)
- 737th Military Police Battalion
- 798th Military Police Battalion (Z of I) (ASF)
- 798th Military Police Battalion (Z of I) (less 2 Composite Companies) (ASF)
- 798th Military Police Battalion (less Detachment) (ASF)
- 798th Military Police Battalion (less Company C)
- 798th Military Police Battalion, HQ & HQ Company & Company C
- 1219th Military Police Company (Aviation)
- 1220th Military Police Company (Aviation)
- 1231st Military Police Company (Aviation)
- 1232nd Military Police Company (Aviation)
- 1233rd Military Police Company (Aviation)
- 1234th Military Police Company (Aviation)
- 1235th Military Police Company (Aviation)
- 1236th Military Police Company (Aviation)
- 1237th Military Police Company (Aviation)
- 1238th Military Police Company (Aviation)
- 1239th Military Police Company (Aviation)
- 1240th Military Police Company (Aviation)
- 1241st Military Police Company (Aviation)
- 1242nd Military Police Company (Aviation)
- 1501st Corps Area Service Unit, Medical Laboratory, Fifth Corps Area
- 1501st Service Command Unit: Fifth Service Command Medical Laboratory
- 1505th Service Command Unit: U.S. Army Finance Office (ASF)
- 1506th Corps Area Service Unit, Fort Benjamin Harrison
- 1506th Service Command Unit: Sub-School for Bakers & Cooks #2 (SOS
- 1511th Service Command Unit: Medical Examination & Induction Board (SOS)
- 1512th Corps Area Service Unit, Motor Pool
- 1512th Service Command Unit: Fifth Service Command Motor Pool (ASF)
- 1516th Service Command Unit: Recruiting & Induction Sub-Station (SOS)
- 1516th Service Command Unit: U.S. Army Recruiting Station (ASF)
- 1526th Service Command Unit: Midwestern Branch, US Disciplinary Barracks
- 1530th Corps Area Service Unit, Headquarters & Headquarters Company
- 1530th Service Command Unit: Station Complement
- 1530th Service Command Unit: Station Complement, WAC Detachment
- 1530th Service Command Unit: Billings General Hospital
- 1530th Service Command Unit: Fifth Service Command Laboratory
- 1530th Service Command Unit: Medical Department Enlisted Technicians School
- 1530th Service Command Unit: Midwestern Branch, US Disciplinary Barracks
- 1530th Service Command Unit: Prisoner of War Camp

1530th Service Command Unit: Statopm Complement (incl WAC personnel)
1530th Service Command Unit: Medical Laboratory
1532nd Service Command Unit: Basic Training Unit (ASF)
1534th Corps Area Service Unit, Reception Center
1534th Service Command Unit: Reception Center (ASF)
1566th Service Command Unit: Sub-School for Bakers & Cooks (ASF)
1584th Service Command Unit: Special Training Unit (ASF)
3548th Service Command Unit: Army Finance School (ASF)
3548th Service Command Unit: Army Finance School, WAC Detachment (ASF)
3552nd Service Command Unit: Casual Company (ASF)
3585th Service Command Unit: WAC Detachment (ASF)
3586th Service Command Unit: Finance Replacement Training Center (ASF)
3586th Service Command Unit: Finance Training Center (ASF)
3586th Service Command Unit: Army Service Forces Training Center (ASF)
3586th Service Command Unit: Army Service Forces Training Center, WAC Det (ASF)

Army Air Forces
Schoen Field (see Schoen Field)

Camp Breckinridge (II)
Morganfield, Kentucky
Sturgis Army Air Field (see Sturgis Army Air Field)
Army Service Forces
Post Photographic Laboratory (ASF)
Provost Marshal Area #6 (I Act)
Provost Marshal Area #8 (I Act)
Tire Collection Center (I Act)
School for Bakers & Cooks (I Act)
25th Field Hospital (SOS)
25th Field Hospital (ASF)
38th Field Hospital (ASF)
112th Army Postal Unit (Colored) (ASF)
113th Army Postal Unit (ASF)
140th Army Postal Unit (ASF)
141st Army Postal Unit (ASF)
176th Army Postal Unit (ASF)
177th Army Postal Unit (ASF)
182nd Station Hospital (250-Bed) (ASF)
196th Station Hospital (100-Bed) (ASF)
197th Station Hospital (100-Bed) (ASF)
198th Station Hospital (100-Bed) (ASF)
199th Station Hospital (100-Bed) (ASF)
225th Army Postal Unit (ASF)
226th Army Postal Unit (ASF)
226th Quartermaster Battalion, HQ & HQ Detachment (ASF)
226th Quartermaster Battalion, HQ & HQ Detachment w/atchd Medical (ASF)
227th Quartermaster Battalion, HQ & HQ Detachment (ASF)
227th Quartermaster Battalion, HQ & HQ Detachment w/atchd Medical (ASF)

- 228th Quartermaser Battalion, HQ & HQ Detachment (ASF)
- 228th Quartermaser Battalion, HQ & HQ Detachment Watchd Medical (ASF)
- 329th Army Postal Unit (ASF)
- 330th Army Postal Unit (ASF)
- 405th Engineer Water Supply Battalion (SOS)
- 541st Army Postal Unit (SOS)
- 570th Quartermaster Service Battalion, HQ & HQ Detachment (Colored) (ASF)
- 983rd Quartermaster Service Company (Colored)
- 984th Quartermaster Service Company (Colored)
- 985th Quartermaster Service Company (Colored)
- 986th Quartermaster Service Company (Colored)
- 589th Quartermaster Battalion, HQ & HQ Detachment w/atchd Med Det (Colored) (ASF)
- 747th Sanitation Company (Medium) (Colored) (ASF)
- 748th Sanitation Company (Medium) (Colored) (ASF)
- 798th Sanitation Company (Medium) (Colored) (ASF)
- 799th Sanitation Company (Medium) (Colored) (ASF)
- 979th Quartermaster Service Company (Colored) (ASF)
- 980th Quartermaster Service Company (Colored) (ASF)
- 981st Quartermaster Service Company (Colored) (ASF)
- 982nd Quartermaster Service Company (Colored) (ASF)
- 1506th Service Command Unit: Sub-School for Bakers & Cooks #5 (SOS)
- 1513th Service Command Unit: Ordnance Service Command Shop C, District #5 (ASF)
- 1538th Service Command Unit: Camp Breckinridge Internment Camp (SOS)
- 1563rd Service Command Unit: Sub-School for Bakers & Cooks (ASF)
- 1570th Service Command Unit: Station Complement
- 1570th Service Command Unit: Station Complement, WAC Detachment #1
- 1570th Service Command Unit: Station Complement, WAC Detachment #2
- 1570th Service Command Unit: Station Complement, WAC Detachment #3
- 1570th Service Command Unit: Prisoner of War Camp
- 1570th Service Command Unit: Station Complement (incl WAC personnel)
- 1570th Service Command Unit: Photographic Laboratory
- 1570th Service Command Unit: Separation Point
- 1572nd Service Command Unit: US Army WAC Recruiting Station
- 3242nd Quartermaster Service Company (ASF)
- 3243rd Quartermaster Service Company (ASF)
- 3244th Quartermaster Service Company (ASF)
- 3245th Quartermaster Service Company (ASF)
- 3246th Quartermaster Service Company (ASF)
- 3247th Quartermaster Service Company (ASF)
- 3248th Quartermaster Service Company (ASF)
- 3249th Quartermaster Service Company (ASF)
- 3556th Service Command Unit: Casual Company (ASF)
- 3562nd Service Command Unit: Station Complement
- 3562nd Service Command Unit: Station Complement, WAC Section #2
- 4332nd Quartermaster Service Company (Colored) (ASF)
- 4333rd Quartermaster Service Company (Colored) (ASF)
- 4334th Quartermaster Service Company (Colored) (ASF)

U.S. Army Installations - World War II:
A Research Reference

4335th Quartermaster Service Company (Colored) (ASF)
4336th Quartermaster Service Company (Colored) (ASF)
4337th Quartermaster Service Company (Colored) (ASF)
4338th Quartermaster Service Company (Colored) (ASF)
4341st Quartermaster Service Company (Colored) (ASF)

Army Ground Forces

2nd Tank Destroyer Brigade, HQ & HQ Company (AGF)
822nd Tank Destroyer Battalion (Towed)
820th Tank Destroyer Battalion (Towed)
7th Antiaircraft Artillery Group, HQ & HQ Battery (AGF)
481st Antiaircraft Artillery Auto Weapons Battalion (Semi Mobile)
9th Tank Battalion
12th Field Artillery Observation Battalion, Battery A (AGF)
13th HQ & HQ Detachment, Special Troops Second Army (AGF)
13th HQ & HQ Detachment, Special Troops Second Army (Type B) (AGF)
13th HQ & HQ Detachment, Special Troops Second Army (Type D) (AGF)
14th HQ & HQ Detachment, Special Troops, Second Army (Type D)
16th Tank Destroyer Group, HQ & HQ Company (AGF)
820th Tank Destroyer Battalion (Towed)
20th Medical Battalion (Separate), HQ & HQ Detachment (AGF)
22nd Ordnance Battalion, HQ & HQ Detachment & Medical Section (AGF)
22nd Ordnance Battalion, HQ & HQ Detachment (AGF)
182nd Ordnance Depot Company
898th Ordnance Heavy Auto Maintenance Company
24th Quartermaster Group, HQ & HQ Detachment (Colored) (AGF)
27th Evacuation Hospital (AGF)
27th Engineer Combat Regiment (AGF)
27th Engineer Regiment Band (AGF)
37th Chemical Decontamination Company (AGF)
30th HQ & HQ Detachment, Special Troops, Second Army
38th Chemical Decontamination Company (Colored) (AGF)
44th Engineer Combat Battalion (AGF)
49th Medical Depot Company (AGF)
50th Medical Depot Company (AGF)
54th Medical Depot Company (AGF)
54th Medical Depot Company, Section 1, Advanced Depot Platoon (AGF)
55th Medical Depot Company, Section 1, Advanced Depot Platoon (AGF)
61st Medical Battalion (AGF)
67th Ordnance Ammunition Battalion, HQ & HQ Detachment & Medical Det (AGF)
35th Infantry Division, Headquarters (less Advance Detachment)
35th Infantry Division, Special Troops, Headquarters
Headquarters Company, 35th Infantry Division
Military Platoon, 35th Infantry Division
35th Infantry Division Band
35th Counter Intelligence Corps Detachment (atchd to HQ Co.)
35th Quartermaster Company
35th Signal Company

- 735th Ordnance Light Maintenance Company
- 134th Infantry Regiment
- 137th Infantry Regiment
- 320th Infantry Regiment
- 35th Division Artillery, HQ & HQ Battery (Mtz) watchd Medical
- 127th Field Artillery Bn (Mtz) (155mm How Tractor-D)
- 161st Field Artillery Bn (Mtz) (105mm How Tractor-D)
- 216th Field Artillery Bn (Mtz) (105mm How Tractor-D)
- 219th Field Artillery Bn (Mtz) (105mm How Tractor-D)
- 60th Engineer Combat Battalion
- 110th Medical Battalion
- 35th Cavalry Reconnaissance Troop, Mechanized
- 194th Photo Interpreter Team (Atchd)
- 75th Infantry Division, Headquarters (AGF)
- 75th Infantry Division, Special Troops, Headquarters
- Headquarters Co, 75th Infantry Divison watchd Med & Chap
- 75th Infantry Division, Military Police Platoon
- 75th Infantry Division Band
- 575th Signal Company
- 775th Ordnance Light Maintenance Company
- 75th Quartermaster Company
- 289th Infantry Regiment
- 290th Infantry Regiment
- 291st Infantry Regiment
- 75th Division Artillery, HQ & HQ Battery (Mtz)
- 730th Field Artillery Battalion (Mtz) (155mm Howitzer, Truck-D)
- 897th Field Artillery Battalion (Mtz) (105mm Howitzer, Truck-D)
- 898th Field Artillery Battalion (Mtz) (105mm Howitzer, Truck-D)
- 899th Field Artillery Battalion (Mtz) (105mm Howitzer, Truck-D)
- 75th Reconnaissance Troop, Mechanized
- 275th Engineer Combat Battalion
- 375th Medical Battalion
- 83rd Infantry Division, Headquarters (AGF)
- 83rd Infantry Division, Special Troops, Headquarters
- Headquarters Company, 83rd Infantry Division watchd Med & Chap
- 83rd Infantry Division, Military Police Platoon
- 83rd Infantry Division Band
- 783rd Ordnance Light Maintenance Company
- 83rd Quartermaster Company
- 83rd Signal Company
- 329th Infantry Regiment
- 330th Infantry Regiment
- 331st Infantry Regiment
- 83rd Division Artillery, HQ & HQ Battery (Motorized)
- 322nd Field Artillery Battalion (Mtz) (105mm Howitzer, Truck-D)
- 323rd Field Artillery Battalion (Mtz) (105mm Howitzer, Truck-D)
- 324th Field Artillery Battalion (Mtz) (155mm Howitzer, Truck-D)

U.S. Army Installations - World War II:
A Research Reference

- 908th Field Artillery Battalion (Mtz) (105mm Howitzer, Truck-D)
- 83rd Reconnaissance Troop, Mechanized
- 308th Engineer Combat Battalion
- 308th Medical Battalion
- 90th Engineer Heavy Ponton Battalion, Co A & HQ Platoon Co B (AGF)
- 95th Evacuation Hospital (Mtz) (AGF)
- 98th Division Artillery Band (AGF)
- 391st Infantry Regiment Band (AGF)
- 98th Infantry Division, Headquarters (AGF)
- 98th Infantry Division, Headquarters Company
- 98th Infantry Division, Military Police Platoon
- 98th Cavalry Reconnaissance Troop
- 98th Infantry Division, Cavalry Reconnaissance Troop
- 98th Signal Company
- 98th Infantry Division, Signal Company
- 389th Infantry Regiment
- 389th Infantry Regiment (less Band)
- 390th Infantry Regiment
- 390th Infantry Regiment (less Band)
- 391st Infantry Regiment
- 98th Division Artillery, Headquarters
- 98th Division Artillery, Headquarters Battery
- 367th Field Artillery Battalion (105mm Howitzer, Truck-D)
- 368th Field Artillery Battalion (105mm Howitzer, Truck-D)
- 369th Field Artillery Battalion (155mm Howitzer, Truck-D)
- 923rd Field Artillery Battalion (105mm Howitzer, Truck-D)
- 98th Quartermaster Company
- 323rd Engineer Battalion
- 323rd Medical Battalion
- 798th Ordnance Light Maintenance Company
- 100th Evacuation Hospital, Semi Mobile (AGF)
- 100th Quartermaster Bakery Battalion, Company D (AGF)
- 101st Portable Surgical Hospital (AGF)
- 102nd Portable Surgical Hospital (AGF)
- 101st Evacuation Hospital, Semi Mobile (AGF)
- 122nd Evacuation Hospital, Semi Mobile (AGF)
- 128th Quartermaster Bakery Company (Colored) (AGF)
- 130th Coast Artillery Battalion (Antiaircraft) (Gun) (Semi Mobile) (AGF)
- 131st Ordnance Medium Maintenance Company (AGF)
- 131st Ordnance Medium Maintenance Company, Field A (AGF)
- 133rd Ordnance Medium Maintenance Company (AGF)
- 136th Evacuation Hospital, Semi Mobile (400 Patient Capacity)
- 138th Quartermaster Battalion, Mobile, HQ & HQ Detachment (AGF)
- 138th Quartermaster Battalion, Mobile, HQ & HQ Detachment watchd Medical (AGF)
- 3799th Quartermaster Truck Company (Troop) (Colored) (AGF)
- 3800th Quartermaster Truck Company (Troop) (Colored) (AGF)

U.S. Army Installations - World War II:
A Research Reference

- 3957th Quartermaster Truck Company (Troop) (Colored) (AGF)
- 3958th Quartermaster Truck Company (Troop) (Colored) (AGF)
- 3959th Quartermaster Truck Company (Troop) (Colored) (AGF)
- 3960th Quartermaster Truck Company (Troop) (Colored) (AGF)
- 3973rd Quartermaster Truck Company (Troop) (Colored) (AGF)
- 149th Chemical Decontamination Company (Colored) (AGF)
- 164th Signal Photo Company – 1 General Assignment Unit (AGF)
- 164th Signal Photo Company – General Assignment Unit #5 (AGF)
- 164th Ordnance Battalion, HQ & HQ Detachment watchd Medical
- 165th Engineer Combat Battalion (AGF)
- 169th Ordnance Ammunition Battalion, HQ & HQ Detachment (Colored) (AGF)
- 169th Quartermaster Battalion, Mobile, HQ & HQ Detachment
- 891st Quartermaster Troop Transport Company (Colored)
- 892nd Quartermaster Troop Transport Company (Colored)
- 893rd Quartermaster Troop Transport Company (Colored)
- 894th Quartermaster Troop Transport Company (Colored)
- 895th Quartermaster Troop Transport Company (Colored)
- 896th Quartermaster Troop Transport Company (Colored)
- 172nd Engineer Combat Battalion (AGF)
- 183rd Medical Battalion, HQ & HQ Detachment (AGF)
- 184th Medical Battalion, HQ & HQ Detachment (AGF)
- 190th Quartermaster Battalion, Mobile, HQ & HQ Detachment watchd Medical
- 193rd Signal Repair Company – 4 Radio Repair Sections (AGF)
- 194th AGF Band (AGF)
- 196th Station Hospital (100-Bed) (ASF)
- 199th AGF Band
- 208th Engineer Combat Battalion (AGF)
- 209th Ordnance Ammunition Company (Colored) (AGF)
- 210th Ordnance Battalion, HQ & HQ Detachment (AGF)
- 377th Ordnance Medium Auto Maintenance Company
- 218th Signal Depot Company (AGF)
- 227th AGF Band (Colored) (AGF)
- 238th Ordnance Ammunition Company (Colored) (AGF)
- 239th Ordnance Ammunition Company (Colored) (AGF)
- 243rd Engineer Combat Battalion (AGF)
- 271st Ordnance Medium Maintenance Company (AGF)
- 284th Engineer Combat Battalion (AGF)
- 286th Ordnance Medium Maintenance Company
- 330th Ordnance Battalion, HQ & HQ Detachment watchd Medical Det (AGF)
- 330th Ordnance Battalion, HQ & HQ Detachment
- 131st Ordnance Heavy Maintenance Company, Field A
- 382nd Ordnance Medium Auto Maintenance Company
- 387th Ordnance Medium Auto Maintenance Company
- 543rd Ordnance Heavy Maintenance Company, Field A
- 335th Quartermaster Depot Company (AGF)
- 339th Quartermaster Depot Company (AGF)
- 341st Medical Regiment (less Band) (AGF)

341st Medical Group, HQ & HQ Detachment (AGF)
20th Medical Battalion (Separate), HQ & HQ Detachment
515th Clearing Company (Medical) (Separate)
516th Clearing Company (Medical) (Separate)
553rd Ambulance Company, Mtr (Medical) (Separate)
554th Ambulance Company, Mtr (Medical) (Separate)
183rd Medical Battalion (Separate), HQ & HQ Detachment
473rd Medical Collecting Company (Separate)
474th Medical Collecting Company (Separate)
475th Medical Collecting Company (Separate)
625th Clearing Company (Medical) (Separate)
184th Medical Battalion (Separate), HQ & HQ Detachment
626th Clearing Company (Medical) (Separate)
350th Ordnance Battalion, HQ & HQ Detachment watchd Medical & Chaplain
370th Infantry Regiment (AGF)
370th Infantry Regiment Band (AGF)
372nd Infantry Regiment (Colored) (AGF)
377th Ordnance Medium Auto Maintenance Company (AGF)
382nd Ordnance Medium Auto Maintenance Company (AGF)
387th Ordnance Medium Maintenance Company (AGF)
391st Infantry Regiment Band (AGF)
406th Quartermaster Truck Company
417th Field Artillery Group (Motorized), HQ & HQ Battery (AGF)
765th Field Artillery Battalion (Mtz) (155mm Howitzer, Tractor-D)
766th Field Artillery Battalion (Mtz) (155mm Howitzer, Tractor-D)
767th Field Artillery Battalion (Mtz) (155mm Howitzer, Tractor-D)
468th Quartermaster Truck Regiment (AGF)
468th Quartermaster Truck Regiment (less 1st Bn 2nd Bn & Co L) (Colored) (AGF)
503rd Quartermaster Car Company (AGF)
505th Military Police Battalion (Fld A) (AGF)
513th Ordnance Heavy Maintenance Company (Field Army) (AGF)
515th Clearing Company (Medical) (Separate) (AGF)
516th Clearing Company (Medical) (Separate) (AGF)
519th Engineer Maintenance Company (AGF)
533rd Quartermaster Railhead Company (AGF)
536th Engineer Light Ponton Company (AGF)
536th Ordnance Heavy Maintenance Company, Tank
538th Engineer Light Ponton Company
551st Engineer Heavy Ponton Battalion
553rd Ambulance Company, Motor (Medical) (Separate) (AGF)
554th Ambulance Company, Motor (Medical) (Separate) (AGF)
556th Quartermaster Railhead Company (AGF)
557th Engineer Combat Battalion (AGF)
557th Quartermaster Railhead Company (AGF)
557th Engineer Heavy Ponton Battalion (AGF)
557th Engineer Heavy Ponton Battalion (less Detachments) (AGF)
572nd Quartermaster Railhead Company (AGF)

588th Engineer Light Ponton Company (AGF)
591st Ordnance Ammunition Company (AGF)
597th Ordnance Ammunition Company (AGF)
597th Ordnance Ammunition Company (Colored) (AGF)
598th Field Artillery Battalion (105mm Howitzer, Truck-D) (AGF)
598th Ordnance Ammunition Company (AGF)
599th Ordnance Ammunition Company (AGF)
599th Ordnance Ammunition Company (Colored) (AGF)
600th Ordnance Ammunition Company (AGF)
600th Ordnance Ammunition Company (Colored) (AGF)
609th Tank Destroyer Battalion (Self-Propelled)
616th Ordnance Ammunition Company (AGF)
618th Ordnance Ammunition Company (AGF)
620th Ordnance Ammunition Company (AGF)
631st Engineer Light Equipment Company (AGF)
666th Clearing Company (Medical) (Separate)
667th Ordnance Ammunition Company (Colored) (AGF)
668th Ordnance Ammunition Company (Colored) (AGF)
676th Ordnance Ammunition Company (Colored)
695th Ordnance Ammunition Company (Colored)
722nd Engineer Depot Company (AGF)
732nd Engineer Depot Company (AGF)
764th Field Artillery Battalion (Motorized) (155mm Howitzer, Tractor-D) (AGF)
765th Field Artillery Battalion (Motorized) (155mm Howitzer, Tractor-D) (AGF)
766th Field Artillery Battalion (Motorized) (155mm Howitzer, Tractor-D) (AGF)
767th Field Artillery Battalion (Motorized) (155mm Howitzer, Tractor-D) (AGF)
789th Ordnance Light Maintenance Company (Infantry Division) (AGF)
807th Tank Destroyer Battalion (Self-Propelled)
817th Tank Destroyer Battalion (Towed) (AGF)
820th Tank Destroyer Battalion (Towed) (AGF)
821st Tank Destroyer Battalion (Towed) (AGF)
822nd Tank Destroyer Battalion (Towed) (AGF)
825th Tank Destroyer Battalion (Towed) (AGF)
884th Quartermaster Gas Supply Company (Colored)
898th Ordnance Heavy Maintenance Company (AGF)
898th Ordnance Heavy Auto Maintenance Company (AGF)
935th Ordnance Heavy Auto Maintenance Company
968th Engineer Maintenance Company (AGF)
969th Ordnance Heavy Auto Maintenance Company (AGF)
979th Engineer Maintenance Company (AGF)
1021st Engineer Treadway Bridge Company
1022nd Engineer Treadway Bridge Company
1105th Engineer Combat Group, HQ & HQ Company (AGF)
27th Engineer Combat Battalion
208th Engineer Combat Battalion
1126th Engineer Combat Group, HQ & HQ Company (AGF)
519th Engineer Maintenance Company

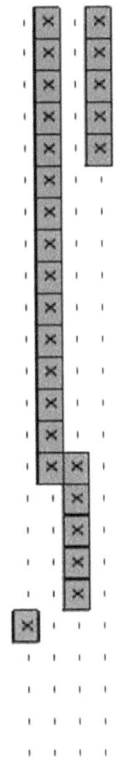

1271st Engineer Combat Battalion
2730th Engineer Light Equipment Company
2731st Engineer Light Equipment Company
1140th Engineer Combat Group, HQ & HQ Company (AGF)
1162nd Engineer Combat Group, HQ & HQ Company (AGF)
1164th Engineer Combat Group, HQ & HQ Company (AGF)
588th Engineer Light Ponton Company
1293rd Engineer Combat Battalion
1479th Engineer Maintenance Company
2732nd Engineer Light Equipment Company
2733rd Engineer Light Equipment Company
1271st Engineer Combat Battalion (AGF)
1280th Engineer Combat Battalion
1293rd Engineer Combat Battalion (AGF)
1479th Engineer Maintenance Company (AGF)
2730th Engineer Light Equipment Company (AGF)
2731st Engineer Light Equipment Company (AGF)
2732nd Engineer Light Equipment Company (AGF)
2733rd Engineer Light Equipment Company (AGF)
3229th Ordnance Depot Company
3416th Quartermaster Truck Company (Colored)
3457th Ordnance Medium Auto Maintenance Company
3480th Ordnance Medium Maintenance Company (Q) (AGF)
3531st Ordnance Auto Maintenance Company (AGF)
3549th Ordnance Medium Auto Maintenance Company
3613th Ordnance Medium Maintenance Company
3673rd Quartermaster Truck Company (Colored)
3674th Quartermaster Truck Company (Colored)
3695th Quartermaster Truck Company (Colored)
3699th Quartermaster Truck Company (Colored)
3799th Quartermaster Truck Company (Troop) (AGF)
3800th Quartermaster Truck Company (Troop) (AGF)
3957th Quartermaster Truck Company (Troop) (Colored) (AGF)
3958th Quartermaster Truck Company (Troop) (AGF)
3959th Quartermaster Truck Company (Troop) (AGF)
3960th Quartermaster Truck Company (Troop) (Colored) (AGF)
3973rd Quartermaster Truck Company (Troop) (AGF)
4027th Quartermaster Truck Company (Colored)
4295th Quartermaster Gas Supply Company (AGF)

Camp Campbell (II)
Clarksville, Tennessee (see note regarding location of Camp Campbell)

Army Service Forces
Fifth Service Command and Signal Corps Photographic Laboratory (SOS)
Provost Marshal Area #9 (I Act)
Post Photographic Hospital
Tire Collection Center (I Act)

School for Bakers & Cooks (I Act)
8th Machine Records Unit (Mobile) (AG) (Type Y) (ASF)
27th Field Hospital (ASF)
28th Field Hospital (ASF)
29th Field Hospital (ASF)
30th Field Hospital (ASF)
40th General Hospital (ASF)
110th General Hospital (ASF)
298th Quartermaster Salvage Repair Company (SOS)
823rd Military Police Company
1506th Service Command Unit: Sub-School for Bakers & Cooks #4 (SOS)
1513th Service Command Unit: Ordnance Service Command Shop (ASF)
1513th Service Command Unit: Ordnance Service Command Shop A, District #4 (ASF)
1539th Service Command Unit: Camp Campbell Internment Camp (SOS)
1565th Service Command Unit: Sub-School for Bakers & Cooks (ASF)
1565th Service Command Unit: School for Bakers & Cooks (ASF)
1572nd Service Command Unit: US Army Recruiting Station (ASF)
1573rd Service Command Unit: US Army WAC Recruiting Station
1580th Service Command Unit: Station Complement
1580th Service Command Unit: Station Complement, WAC Detachment
1580th Service Command Unit: Prisoner of War Camp
1580th Service Command Unit: Station Complement (includes WAC personnel)
1580th Service Command Unit: Photographic Laboratory
1580th Service Command Unit: Separation Point
1580th Service Command Unit: Sub-School for Bakers & Cooks
3557th Service Command Unit: Casual Company (ASF)
3563rd Service Command Unit: WAC Station Complement (ASF)

Army Ground Forces

IV Armored Corps, HQ & HQ Company (AGF)
XVIII Corps, HQ & HQ Company watchd Medical
XVIII Corps, Military Police Platoon
63rd Order of Battle Team (Atchd)
64th Order of Battle Team (Atchd)
181st Photo Interpreter Team (Atchd)
182nd Photo Interpreter Team (Atchd)
183rd Photo Interpreter Team (Atchd)
218th Counter Intelligence Corps Detachment (Atchd)
XVIII Corps Artillery (Mtz), HQ & HQ Battery
XX Corps, HQ & HQ Company (AGF)
XX Corps Military Police Platoon
XXII Corps, HQ & HQ Company (AGF)
XXII Corps, Military Police Platoon
XXII Corps Artillery (MTZ), HQ & HQ Battery (AGF)
8th Machine Records Unit (Mobile) (AG) (Type Y) (AGF)
9th Armored Regiment Band (AGF)
20th Armored Regiment Band (AGF)
43rd Armored Regiment Band (AGF)

- 44th Armored Regiment Band (AGF)
- 5th Armored Signal Battalion (AGF)
- 5th Infantry Division, Headquarters
  - HQ Company, 5th Infantry Division watchd Medical & Chaplain
  - 5th Counter Intelligence Corps Detachment (Atchd)
  - 5th Quartermaster Company
  - 5th Signal Company
  - 705th Ordnance Light Maintenance Company
  - Military Police Platoon, 5th Infantry Division
  - 5th Infantry Division Band
- 2nd Infantry Regiment
- 10th Infantry Regiment
- 11th Infantry Regiment
- 5th Division Artillery, HQ & HQ Battery
  - 19th Field Artillery Battalion (Mtz) (105mm Howitzer, Truck-D)
  - 21st Field Artillery Battalion (Mtz) (155mm Howitzer, Truck-D)
  - 46th Field Artillery Battalion (Mtz) (105mm Howitzer, Truck-D)
  - 50th Field Artillery Battalion (Mtz) (105mm Howitzer, Truck-D)
- 7th Engineer Combat Battalion
- 5th Medical Battalion
- 5th Cavalry Reconnaissance Troop, Mechanized
- 72nd Order of Battle Team (Atchd)
- 189th Photo Interpreter Team (Atchd)
- 8th Armored Division, HQ & HQ Company (less Detachments) (AGF)
- 8th Armored Division Service Company
- 148th Armored Signal Company
- 88th Armored Reconnaissance Battalion
- 36th Armored Regiment
- 80th Armored Regiment
- 398th Armored Field Artillery Battalion
- 399th Armored Field Artillery Battalion
- 405th Armored Field Artillery Battalion
- 53rd Armored Engineer Battalion
- 49th Armored Infantry Regiment
- Division Trains, HQ & HQ Company
  - 8th Armored Division, Trains, Supply Battalion
  - 8th Armored Division, Trains, Maintenance Battalion
- 78th Armored Medical Battalion
- 9th Tank Group, HQ & HQ Detachment
  - 701st Tank Battalion (Medium)
  - 702nd Tank Battalion (Medium)
  - 782nd Tank Battalion (Light)
- 11th Tank Group, HQ & HQ Detachment (AGF)
  - 701st Tank Battalion (Medium)
  - 702nd Tank Battalion (Medium)
  - 782nd Tank Battalion (Medium)

- 12th Armored Division, HQ & HQ Company (AGF)
- 12th Armored Division, Service Company
- 152nd Armored Signal Company
- 92nd Armored Reconnaissance Battalion
- 43rd Armored Regiment
- 44th Armored Regiment
- 493rd Armored Field Artillery Battalion
- 494th Armored Field Artillery Battalion
- 495th Armored Field Artillery Battalion
- 119th Armored Engineer Battalion
- 56th Armored Infantry Regiment
- Division Trains, HQ & HQ Company
- Division Trains, Maintenance Battalion
- Division Trains, Support Battalion
- 82nd Armored Medical Battalion
- 13th HQ & HQ Detachment, Special Troops, Second Army (Type D)
- 13th HQ & HQ Detachment, Special Troops, Second Army
- 14th HQ & HQ Detachment, Special Troops, Second Army (AGF)
- 14th Cavalry Group, Hq & HQ Troop
- 14th Armored Division, Headquarters
- 14th Armored Division, Headquarters Company
- 14th Armored Division, Headquarters Company, Combat Command "A"
- 14th Armored Division, Headquarters Company, Combat Command "B"
- 14th Armored Division, Reserve Command
- 18th Chemical Maintenance Company (AGF)
- 154th Armored Signal Company
- 94th Cavalry Reconnaissance Squadron, Mechanized
- 25th Tank Battalion
- 47th Tank Battalion
- 48th Tank Battalion
- 19th Armored Infantry Battalion
- 62nd Armored Infantry Battalion
- 68th Armored Infantry Battalion
- 14th Armored Division Artillery, HQ, Division Artillery Command
- 499th Armored Field Artillery Battalion
- 500th Armored Field Artillery Battalion
- 501st Armored Field Artillery Battalion
- 125th Armored Engineer Battalion
- 14th Armored Division Trains, HQ & HQ Company
- 14th Armored Division Band
- 136th Ordnance Maintenance Battalion (Armored Division)
- 84th Medical Battalion, Armored
- Military Police Platoon, 14th Armored Division
- 14th HQ & HQ Detachment, Special Troops, Second Army (Type A) (AGF)
- 14th Tank Destroyer Group, HQ & HQ Company (AGF)
- 692nd Tank Destroyer Battalion (Towed)
- 14th Armored Group, HQ & HQ Company (AGF)

- 14th HQ & HQ Detachment, Special Troops... Second Army (Type A) (AGF)
- 14th HQ & HQ Detachment, Special Troops... Second Army (Type D) (AGF)
- 19th Ordnance Medium Maintenance Company
- 19th Cavalry Reconnaissance Squadron, Mechanized
- 20th Armored Division, HQ & HQ Company
- 20th Armored Division, Service Company
- 160th Armored Signal Company
- 100th Armored Reconnaissance Battalion
- 9th Armored Regiment
- 20th Armored Regiment
- 412th Armored Field Artillery Battalion
- 413th Armored Field Artillery Battalion
- 414th Armored Field Artillery Battalion
- 480th Armored Infantry Regiment
- 220th Armored Engineer Battalion
- Division Trains, HQ & HQ Company
- Division Trains, Maintenance Battalion
- Division Trains, Support Battalion
- 220th Armored Medical Battalion
- 20th Armored Division, Headquarters (AGF)
- 20th Armored Division, Headquarters Company
- 20th Armored Division, HQ & HQ Company, Combat Command "A"
- 20th Armored Division, HQ & HQ Company, Combat Command "B"
- 20th Armored Division, Reserve Command (HQ)
- 160th Armored Signal Company
- 33rd Cavalry Reconnaissance Squadron, Mechanized
- 9th Tank Battalion
- 20th Tank Battalion
- 27th Tank Battalion
- 8th Armored Infantry Battalion
- 65th Armored Infantry Battalion
- 70th Armored Infantry Battalion
- 20th Armored Division Artillery, HQ, Division Artillery Command
- 412th Armored Field Artillery Battalion
- 413th Armored Field Artillery Battalion
- 414th Armored Field Artillery Battalion
- 220th Armored Engineer Battalion
- 20th Armored Division Trains, HQ & HQ Company
- 20th Armored Division Band
- 138th Ordnance Maintenance Battalion (Armored Division)
- 220th Medium Battalion, Armored
- Military Police Platoon, 20th Armored Division
- 520th Counter Intelligence Corps Detachment (Atchd to HQ Company)
- 22nd Armored Division, Maintenance Battalion (AGF)
- 59th Chemical Maintenance Company (AGF)
- 208th Quartermaster Gas Supply Bn HQ & HQ Det & Medical Det (C
- 312th Ordnance Battalion, HQ & HQ Det & Atchd Medical Det

- 333rd Ordnance Motor Transport Company (Q)
- 452nd Ordnance Evacuation Company
- 482nd Ordnance Evacuation Company
- 859th Ordnance Heavy Auto Maintenance Company
- 23rd Quartermaster Group, HQ & HQ Detachment (AGF)
- 24th Quartermaster Group, HQ & HQ Detachment (Colored)
- 26th Infantry Division, Headquarters (AGF)
- 26th Infantry Division, Special Troops, Headquarters
- HQ Company, 26th Infantry Division, Watchd Medical & Chap
- Military Police Platoon
- 26th Infantry Division Band
- 726th Ordnance Light Maintenance Company
- 26th Quartermaster Company
- 39th Signal Company
- 101st Infantry Regiment
- 104th Infantry Regiment
- 328th Infantry Regiment
- 26th Division Artillery, HQ & HQ Battery (Motorized)
- 101st Field Artillery Battalion (Mtz) (105mm Howitzer, Truck-D)
- 102nd Field Artillery Battalion (Mtz) (105mm Howitzer, Truck-D)
- 180th Field Artillery Battalion (Mtz) (155mm Howitzer, Truck-D)
- 263rd Field Artillery Battalion (Mtz) (105mm Howitzer, Truck-D)
- 26th Reconnaissance Troop, Mechanized
- 101st Engineer Combat Battalion
- 114th Medical Battalion
- 31st Antiaircraft Artillery Group, HQ & HQ Battery (AGF)
- 40th Signal Construction Battalion (Colored) (AGF)
- 58th Signal Repair Company
- 59th Chemical Maintenance Company (AGF)
- 62nd Ordnance Group, HQ & HQ Detachment (AGF)
- 86th Chemical Mortar Battalion
- 101st Cavalry Reconnaissance Squadron, Mechanized
- 101st Quartermaster Bakery Company, 4th Platoon (AGF)
- 103rd Signal Light Construction Battalion
- 106th AGF Band
- 108th Evacuation Hospital, Semi Mobile (AGF)
- 112th Antiaircraft Artillery Group, HQ & HQ Battery (AGF)
- 116th Cavalry Reconnaissance Squadron, Mechanized
- 118th Cavalry Reconnaissance Squadron, Mechanized
- 125th Evacuation Hospital, Semi-Mobile (400 patient capacity)
- 126th Quartermaster Bakery Company (less 1 Platoon) (AGF)
- 128th Quartermaster Bakery Company (Colored) (AGF)
- 128th Ordnance Medium Maintenance Company (AGF)
- 130th Quartermaster Bakery Company (Colored) (AGF)
- 142nd Quartermaster Battalion, Mobile, HQ & HQ Detachment Watchd Medical
- 143rd Quartermaster Battalion, Mobile, HQ & HQ Det. Watchd Medical (AGF)
- 143rd Quartermaster Battalion, Mobile, HQ & HQ Detachment (AGF)

- 3714th Quartermaster Truck Company (Troop) (Colored)
- 4119th Quartermaster Truck Company (Troop) (Colored)
- 151st Ordnance Battalion, HQ & HQ Detachment (AGF)
- 223rd Ordnance Heavy Maintenance Company, Field A
- 440th Ordnance Heavy Auto Maintenance Company
- 674th Ordnance Ammunition Company (Colored)
- 929th Ordnance Heavy Auto Maintenance Company
- 3235th Ordnance Depot Company
- 157th Ordnance Battalion, HQ & HQ Detachment Watchd Medical & Chaplain
- 164th Signal Photo Lab -General Assignment Unit #2 (AGF)
- 185th Engineer Combat Battalion
- 195th Quartermaster Battalion, Mobile, HQ & HQ Detachment (AGF)
- 223rd Ordnance Heavy Maintenance Company, Tank (less 1 Platoon) (AGF)
- 223rd Ordnance Heavy Maintenance Company, Tank (AGF)
- 224th Army Ground Forces Band (AGF)
- 253rd Armored Field Artillery Battalion (AGF)
- 253rd Engineer Combat Battalion (AGF)
- 255th Engineer Combat Battalion
- 259th Quartermaster Railhead Company (AGF)
- 285th Engineer Combat Battalion (AGF)
- 290th Field Artillery Observation Battalion (AGF)
- 295th Ordnance Medium Maintenance Company (AGF)
- 296th Ordnance Medium Maintenance Company (AGF)
- 302nd Ordnance Medium Maintenance Company
- 303rd Signal Operations Battalion
- 310th Signal Operations Battalion (AGF)
- 312th Ordnance Battalion, HQ & HQ Detachment (AGF)
- 333rd Ordnance Depot Company
- 482nd Ordnance Evacuation Company
- 329th Ordnance Battalion, HQ & HQ Detachment (AGF)
- 183rd Ordnance Depot Company
- 562nd Ordnance Heavy Maintenance Company, Tank
- 334th Quartermaster Depot Company (AGF)
- 334th Ordnance Medium Auto Maintenance Company
- 336th Ordnance Battalion, HQ & HQ Detachment Watchd Medical & Chaplain
- 340th Quartermaster Depot Company (AGF)
- 341st Quartermaster Depot Company (AGF)
- 346th Ordnance Battalion, HQ & HQ Detachment (AGF)
- 674th Ordnance Ammunition Company (Colored)
- 929th Ordnance Heavy Auto Maintenance Company
- 349th Ordnance Depot Company (AGF)
- 365th Engineer General Service Regiment (less Band) (AGF)
- 396th Ordnance Medium Auto Maintenance Company (AGF)
- 405th Field Artillery Group (Mtz), HQ & HQ Battery (AGF)
- 274th Armored Field Artillery Battalion
- 275th Armored Field Artillery Battalion
- 276th Armored Field Artillery Battalion

- 417th Ordnance Evacuation Company (AGF)
- 440th Ordnance Heavy Auto Maintenance Company (AGF)
- 466th Ordnance Evacuation Company (AGF)
- 467th Ordnance Evacuation Company (AGF)
- 474th Antiaircraft Artillery Auto Weapons Battalion (Self-Propelled) (AGF)
- 475th Ordnance Evacuation Company (AGF)
- 482nd Ordnance Evacuation Company (AGF)
- 489th Quartermaster Depot Company (AGF)
- 500th Engineer Light Ponton Company
- 503rd Ordnance Heavy Maintenance Company (AGF)
- 519th Engineer Maintenance Company
- 527th Quartermaster Railhead Company (AGF)
- 533rd Quartermaster Railhead Company (AGF)
- 537th Armored Infantry Battalion (AGF)
- 539th Engineer Light Ponton Company
- 554th Engineer Heavy Ponton Company
- 562nd Quartermaster Service Battalion (Colored) (AGF)
- 562nd Quartermaster Battalion, HQ & HQ Det Watchd Medical Det (Colored) (AGF)
- 3216th Quartermaster Service Company (Colored)
- 3217th Quartermaster Service Company (Colored)
- 3218th Quartermaster Service Company (Colored)
- 3219th Quartermaster Service Company (Colored)
- 565th Ordnance Heavy Maintenance Company, Tank
- 571st Antiaircraft Artillery Auto Weapons Battalion (Self-Propelled) (AGF)
- 579th Antiaircraft Artillery Auto Weapons Battalion (Self-Propelled) (AGF)
- 604th Engineer Camouflage Battalion (AGF)
- 604th Engineer Camouflage Battalion – less Company A (AGF)
- 615th Quartermaster Depot Company (Supply) (AGF)
- 616th Quartermaster Depot Company (Supply) (AGF)
- 629th Engineer Light Equipment Company (less Detachment)
- 630th Engineer Light Equipment Company
- 635th Quartermaster Laundry Company (Semi Mobile) (AGF)
- 654th Ordnance Ammunition Company (Colored)
- 655th Tank Destroyer, Company B (AGF)
- 656th Tank Destroyer Battalion (Self-Propelled)
- 657th Engineer Topographic Battalion (A) (less 1 Platoon)
- 669th Quartermaster Truck Company (Colored)
- 673rd Engineer Topographic Company, Corps (AGF)
- 674th Ordnance Ammunition Company (Colored) (AGF)
- 692nd Tank Destroyer Battalion (Towed) (AGF)
- 701st Tank Battalion (Medium) (AGF)
- 702nd Tank Battalion (Medium) (AGF)
- 718th Tank Battalion
- 781st Tank Battalion
- 788th Tank Battalion (AGF)
- 805th Replacement Battalion
- 840th Ordnance Depot Company

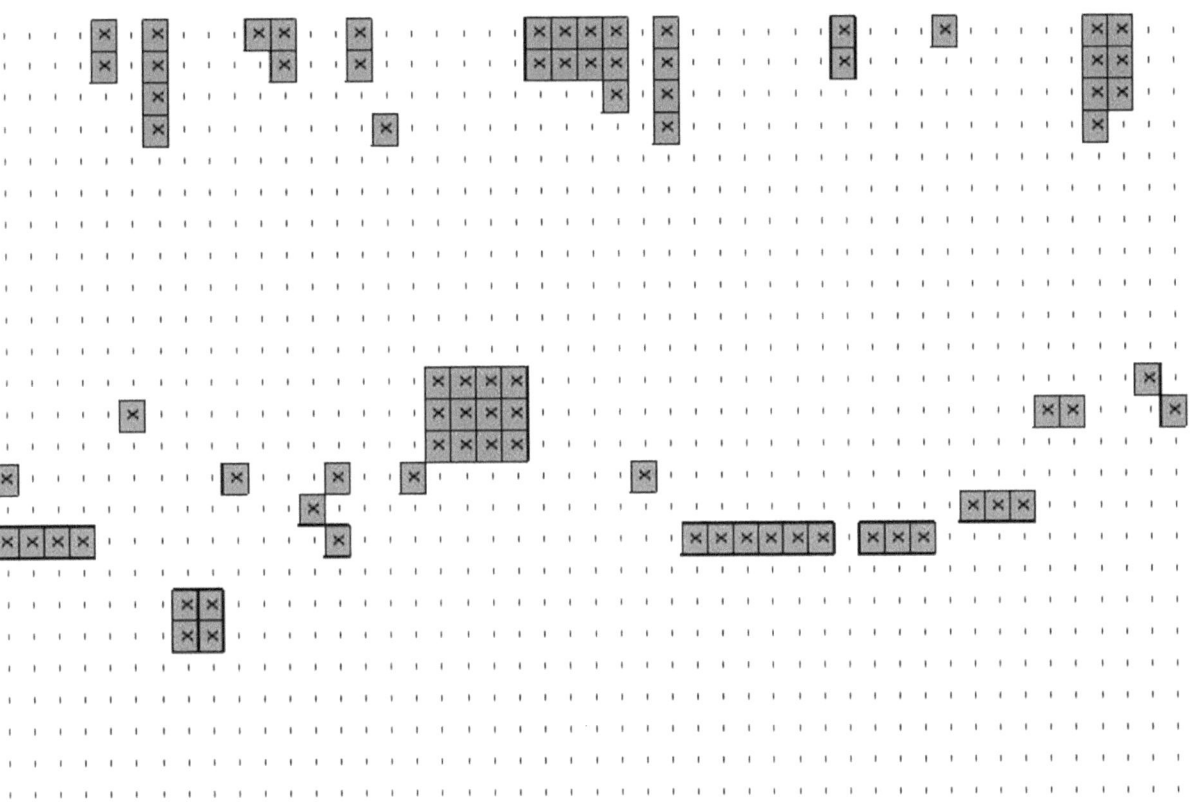

3959th Quartermaster Truck Company (Troop) (Colored) (AGF)
3973rd Quartermaster Truck Company (Troop) (Colored) (AGF)
4032th Quartermaster Truck Company (Troop) (Colored) (AGF)
4225th Quartermaster Truck Company (Troop) (Colored) (AGF)
4288th Quartermaster Railhead Company
4295th Quartermaster Gas Supply Company (Colored)

Fort Knox (II)
Fort Knox, Kentucky
Godman Field (see Godman Field)
Army Service Forces
ASF Regional Hospital (I Act)
Ordnance Officer Replacement Pool (ASF)
Provost Marshal General Area #7 (I Act)
Rehabilitation Center (I Act)
School for Bakers & Cooks
  Sub-School for Bakers & Cooks
Tire Collection Center (I Act)
4th Italian Ordnance Medium Auto Maintenance Company
5th Italian Ordnance Medium Auto Maintenance Company
6th Italian Ordnance Medium Auto Maintenance Company
7th Italian Ordnance Medium Auto Maintenance Company
16th Army Postal Unit
17th Army Postal Unit
26th Medical Depot Company (ASF)
32nd Medical Depot Company (ASF)
52nd Army Postal Unit (APO, GA) (asf)
53rd Army Postal Unit (APO, GA) (asf)
105th Station Hospital (SOS)
111th General Hospital
136th Italian Quartermaster Service Company (less 2nd Platoon)
136th Italian Quartermaster Service Company (less 2nd Platoon & 6 EM from HQ Co.)
137th Italian Quartermaster Service Company (less 2nd Platoon)
137th Italian Quartermaster Service Company (less 2nd Platoon & 6 EM from HQ Co.)
140th General Hospital (ASF)
146th Army Postal Unit (ASF)
147th Army Postal Unit (ASF)
148th Station Hospital (50-Bed) (ASF)
232nd Army Postal Unit (ASF)
233rd Army Postal Unit (ASF)
274th Station Hospital (250-Bed) (ASF)
275th Station Hospital (250-Bed) (ASF)
275th Station Hospital (750-Bed) (ASF)
311th Italian Quartermaster Battalion, HQ & HQ Detachment
311th Italian Quartermaster Battalion, HQ & HQ Detachment w/ atchd Medical
375th Engineer General Service Regiment (less Band) (Colored) (ASF)
377th Engineer General Service Regiment (less Band) (Colored) (ASF)

Add AAF, ASF, POW to Jan 42

- 540th Army Postal Unit (SOS)
- 556th Ordnance Heavy Maintenance Company (Truck) (ASF)
- 573rd Army Postal Unit (ASF)
- 589th Army Postal Unit (ASF)
- 729th Military Police Battalion (ZI), Company C (ASF)
- 729th Military Police Battalion (ZI), Companies C & D (ASF)
- 729th Military Police Battalion (ZI), Company D (ASF)
- 769th Engineer Dump Truck Company (Colored) (ASF)
- 798th Military Police Battalion, Company C
- 798th Military Police Battalion, Company B
- 1506th Service Command Unit: Fifth Service Command School for Bakers & Cooks, HQ
- 1506th Service Command Unit: Fifth Service Command School for Bakers & Cooks
- 1506th Service Command Unit: Sub-School for Bakers & Cooks #1 (SOS)
- 1506th Service Command Unit: Sub-School for Bakers & Cooks #8 (SOS)
- 1513th Corps Area Service Unit, Replacement Center
- 1513th Service Command Unit: Ordnance Service Command Shop A, District #1 (ASF)
- 1516th Service Command Unit: US Army Recruiting Station (ASF)
- 1536th Service Command Unit: Detention & Rehabilitation Center
- 1536th Service Command Unit: Rehabilitation Center
- 1536th Service Command Unit: Rehabilitation Center (Station Complement) (ASF)
- 1550th Corps Area Service Unit, Station Complement
- 1550th Service Command Unit: Station Complement
- 1550th Service Command Unit: Station Complement, WAC Detachment
- 1550th Service Command Unit: Station Complement, WAC Detachment (Colored)
- 1550th Service Command Unit: ASF Regional Hospital
- 1550th Service Command Unit: Bakers & Cooks School
- 1550th Service Command Unit: Prisoner of War Camp
- 1550th Service Command Unit: Photographic Laboratory
- 1550th Service Command Unit: Rehabilitation Center
- 1550th Service Command Unit: Station Complement (includes WAC personnel)
- 1550th Service Command Unit: Separation Point
- 1572nd Service Command Unit: US Army WAC Recruiting Station
- 3554th Service Command Unit: Casual Company (ASF)
- 3564th Service Command Unit: WAC Detachment (ASF)
- 9929th Technical Service Unit SGO: Armored Medical Research Lab (incl WAC prsnl)

Army Ground Forces
- Armored Force, HQ
- Armored Force, HQ & HQ Company
- Armored Force School
- Armored Force Replacement Training Center
- Armored Center, HQ & HQ Company
- Armored Command, HQ & HQ Company
- Armored Force Board
- Armored Force Officer Candidate School (AGF)
- Armored Board
- Armored Replacement Training Center
- Armored Officer Replacement Pool

Armored Replacement Units

- 1st Armored Force Replacement Training Group
  - 1st Armored Force Replacement Training Battalion
  - 2nd Armored Force Replacement Training Battalion
  - 3rd Armored Force Replacement Training Battalion
  - 5th Armored Force Replacement Training Battalion
  - 8th Armored Force Replacement Training Battalion
- 1st Armored Replacement Group
  - 1st Armored Replacement Battalion
  - 2nd Armored Replacement Battalion
  - 3rd Armored Replacement Battalion
  - 4th Armored Replacement Battalion
  - 5th Armored Replacement Battalion
  - 6th Armored Replacement Battalion
- 2nd Armored Force Replacement Training Group
  - 6th Armored Force Replacement Training Battalion
  - 7th Armored Force Replacement Training Battalion
  - 9th Armored Force Replacement Training Battalion
  - 10th Armored Force Replacement Training Battalion
  - 11th Armored Force Replacement Training Battalion
  - 12th Armored Force Replacement Training Battalion
- 2nd Armored Replacement Group
  - 7th Armored Replacement Battalion
  - 8th Armored Replacement Battalion
  - 9th Armored Replacement Battalion
  - 10th Armored Replacement Battalion
  - 11th Armored Replacement Battalion
  - 12th Armored Replacement Battalion
- 3rd Armored Force Replacement Training Group
  - 4th Armored Force Replacement Training Battalion, Cos A & B (Col...)
  - 13th Armored Force Replacement Training Battalion
  - 15th Armored Force Replacement Training Battalion
  - 16th Armored Force Replacement Training Battalion
  - 17th Armored Force Replacement Training Battalion
  - 18th Armored Force Replacement Training Battalion
  - 19th Armored Force Replacement Training Battalion
  - 20th Armored Force Replacement Training Battalion
  - 21st Armored Force Replacement Training Battalion
- 3rd Armored Replacement Group
  - 13th Armored Replacement Battalion
  - 14th Armored Replacement Battalion
  - 15th Armored Replacement Battalion
  - 16th Armored Replacement Battalion
  - 17th Armored Replacement Battalion
  - 18th Armored Replacement Battalion
- 1st Armored Replacement Regiment
  - 1st Armored Replacement Battalion
  - 2nd Armored Replacement Battalion

Unit	Strength
3rd Armored Replacement Battalion	xxxxxxxxxxxxxx
4th Armored Replacement Battalion	xxxxxxxxxxxxxx
13th Armored Replacement Battalion	
14th Armored Replacement Battalion	
15th Armored Replacement Battalion	
16th Armored Replacement Battalion	
17th Armored Replacement Battalion	
18th Armored Replacement Battalion	
2nd Armored Replacement Regiment	
5th Armored Replacement Battalion	xxxxxxxxxxxxxxxx
6th Armored Replacement Battalion	xxxxxxxxxxxxxxxx
7th Armored Replacement Battalion	xxxxxxxxxxxxxxxx
8th Armored Replacement Battalion	xxxxxxxxxxxxxxxx
9th Armored Replacement Battalion	xxxxxxxxxxxxx
10th Armored Replacement Battalion	xxxxxxxxxxxxxxxx
11th Armored Replacement Battalion	xxxxxxxxxxxxxxxx
12th Armored Replacement Battalion	xxxxxxxxxxxxxxxx
3rd Armored Replacement Regiment	
1st Armored Replacement Battalion	xxxxxxxxxxxxxxxx
2nd Armored Replacement Battalion	
3rd Armored Replacement Battalion	
4th Armored Replacement Battalion	
5th Armored Replacement Battalion	
6th Armored Replacement Battalion	
9th Armored Replacement Battalion	xxxxxxxxxxxxxxxx
10th Armored Replacement Battalion	xxxxxxxxxxxxxxxx
11th Armored Replacement Battalion	xxxxxxxxxxxxxxxx
12th Armored Replacement Battalion	xxxxxxxxxxxxxxxx
4th Armored Replacement Regiment	
12th Armored Replacement Battalion	xxxxxxxxxxxxxxxx
13th Armored Replacement Battalion	xxxx
14th Armored Replacement Battalion	xxxx
15th Armored Replacement Battalion	
5th Armored Replacement Regiment	xxxxxxxxxxxxxxxx
16th Armored Replacement Battalion	xxxxxxxxxxxxxxxx
17th Armored Replacement Battalion	xxxxxxxxxxxxxxxx
18th Armored Replacement Battalion	xxxxxxxxxxxxxxxx
19th Armored Replacement Battalion	xxxxxxxxxxxxxxxx
20th Armored Special Training Regiment	xxxxxxxxxxxxxxxx
20th Armored Special Training Battalion	xxxxxxxxxxxxxxxx
21st Armored Special Training Battalion	xxxxxxxxxxxxxxxx
22nd Armored Special Training Battalion	xxxxxxxxxxxxxxxx
23rd Armored Special Training Battalion	xxxxxxxxxxxxxxxx
24th Armored Special Training Battalion	xxxxxxxxxxxxxxxx
25th Armored Special Training Battalion	xxxxxxxxxxxxxxxx
26th Armored Special Training Battalion	xxxxxxxxxxxxxxxx
27th Armored Special Training Battalion	xxxxxxxxxxxxxxxx

- 1st Advanced Armored Training Battalion
- 2nd Advanced Armored Training Battalion
- 19th Special Training Battalion
- The Battle Training Group
- WAC Detachment
- Armored School
  - Armored Center School, Demonstration Regiment
  - Armored Medical Research Laboratory (IV Act)
  - Armored Officer Candidate School
  - Armored Force Training Group
  - Provisional Tank Company
  - WAC Detachment
  - School Troops, HQ & HQ Detachment
  - 1st Tank Training Detachment
  - 2nd Tank Training Detachment
  - 3rd Tank Training Detachment
  - Antiaircraft Artillery Searchlight Training Detachment
  - Infantry-Engineer Training Detachment
  - Reconnaissance Section
  - Medical Detachment
  - Motor Pool Detachment
  - Wheeled Vehicle Section, Motor Pool Detachment
  - Tank Park Section, Motor Pool Detachment
  - Tank Destroyer-Field Artillery Training Detachment
  - Tank Training Detachment
  - Range Detachment
  - Reconnaissance Training Detachment
- Army Ground Forces Board No. 2
  - Amphibious Section (less Amphibious Equipt Branch)
  - Automotive Section
  - Engineer Section (less Engineer Equipt Branch)
  - Maintenance & Supply Section
  - Weapons & Fire Control Section
- Band (242nd Coast Artillery Regiment) (AGF)
- Armored Force Replacement Training Center #1 Band (AGF)
- Armored Force Replacement Training Center #2 Band (AGF)
- 36th Armored Regiment Band
- 80th Armored Regiment Band
- Field Printing Plant
- Armored Corps Replacement Training Center
- Armored Force Replacement Training Center (AGF)
- 1st Armored Replacement Training Center
  - 1st Armored Training Battalion
  - 2nd Armored Training Battalion
  - 3rd Armored Training Battalion
  - 4th Armored Training Battalion
  - 5th Armored Training Battalion

- 6th Armored Training Battalion
- 7th Armored Training Battalion
- 8th Armored Training Battalion
- 9th Armored Training Battalion
- 10th Armored Training Battalion
- 11th Armored Training Battalion
- 12th Armored Training Battalion
- 13th Armored Training Battalion
- 14th Armored Training Battalion
- 15th Armored Training Battalion
- 16th Armored Training Battalion
- 17th Armored Training Battalion
- 18th Armored Training Battalion
- 1st Armored Training Group
- 2nd Armored Training Group
- 3rd Armored Training Group
- I Army Corps, HQ & HQ Company
- I Armored Corps, HQ & HQ Company
- 1st Tank Group, HQ & HQ Company
- 1st Armored Division
- 1st Armored Division, HQ & HQ Company
- 1st Armored Division, Service Company
- 1st Armored Division Train, Headquarters Detachment
- 1st Armored Division Train, Supply Battalion
- 1st Armored Division, Maintenance Battalion
- 1st Signal Battalion (Armored)
- 1st HQ & HQ Detachment, Special Troops, Armored Force (AGF)
- 1st HQ & HQ Detachment, Special Troops, Armored Command (AGF)
- 1st Armored Group (AGF)
- 526th Armored Infantry Battalion
- 527th Armored Infantry Battalion
- 528th Armored Infantry Battalion
- 8th Tank Group, HQ & HQ Detachment (AGF)
- 740th Tank Battalion (Medium)
- 750th Tank Battalion (Medium)
- 781st Tank Battalion (Light)
- 785th Tank Battalion (Light)
- 10th Armored Group, HQ & HQ Company (AGF)
- 10th Reconnaissance Troop, Mechanized (Cavalry) (AGF)
- 10th Cavalry Reconnaissance Troop, Mechanized (AGF)
- 192nd Tank Battalion
- 72nd Field Artillery Brigade, HQ & HQ Battery
- 119th Field Artillery Regiment (155mm Gun)
- 177th Field Artillery Regiment (155mm Howitzer)
- 182nd Field Artillery Regiment (155mm Howitzer)

- 1st Division
- 1st Armored Regiment

- 1st Armored Regimetal Band
- 5th Division
- 5th Armored Division, HQ & HQ Company
- 5th Armored Division, Service Company
- 5th Armored Division Train, Headquarters Detachment
- 5th Armored Division Train, Maintenance Battalion
- 5th Armored Division Train, Supply Battalion
- 6th Infantry Regiment
- 6th Surgical Hospital
- 6th Armored Division, HQ & HQ Company
- 6th Armored Division, Service Company
- 6th Armored Division Train, Headquarters Detachment
- 6th Armored Division Train, Maintenance Battalion
- 6th Armored Division Train, Supply Battalion
- 10th Evacuation Hospital
- 12th Observation Squadron
- 13th Armored Regiment
- 13th Armored Regimental Band
- 13th Quartermaster Battalion
- 14th Armored Group, HQ & HQ Company
- 16th Armored Engineer Battalion
- 16th Engineer Battalion
- 17th Armored Group, HQ & HQ Company (AGF)
- 527th Armored Infantry Battalion
- 785th Tank Battalion (Light)
- 18th Tank Desstroyer Group, HQ & HQ Company
- 19th Quartermaster Battalion
- 20th Armored Division, Support Battalion (AGF)
- 21st Ordnance Battalion (Armored)
- 22nd Armored Engineer Battalion
- 22nd Tank Desstroyer Group, HQ & HQ Company
- 22nd Engineer Battalion (Armored)
- 22nd Ordnance Battalion
- 22nd Quartermaster Regiment, Company K
- 22nd Quartermaster Regiment, Cos K & L (Truck) (Colored)
- 25th Armored Engineer Battalion
- 26th Ordnance Battalion, HQ & HQ Detachment & Medical Detachment
- 26th Ordnance Battalion, HQ & HQ Detachment
- 169th Ordnance Depot Company
- 199th Ordnance Depot Company
- 392nd Ordnance Medium Maintenance Company
- 485th Ordnance Evacuation Company (AGF)
- 528th Ordnance Heavy Maintenance Company (Truck)
- 538th Ordnance Heavy Maintenance Company (Truck)
- 553rd Ordnance Heavy Maintenance Company (Truck)
- 554th Ordnance Heavy Maintenance Company (Truck)

- 556th Ordnance Heavy Maintenance Company (Truck)
- 565th Ordnance Heavy Maintenance Company (Tank)
- 566th Ordnance Heavy Maintenance Company (Tank)
- 619th Ordnance Ammunition Company (Colored)
- 841st Ordnance Depot Company
- 27th Armored Field Artillery Battalion (105mm Howitzer, Armored)
- 28th Airborne Tank Battalion (AGF)
- 31st Armored Regimental Band
- 34th Armored Regiment (less Band)
- 38th Ordnance Company (Medium Maintenance) (Attached)
- 44th Engineer Combat Battalion, Company B (AGF)
- 46th Armored Infantry Regiment
- 46th Infantry Regiment
- 47th Medical Battalion (Armored)
- 47th Armored Signal Company
- 50th Armored Infantry Battalion
- 53rd Ordnance Company (Ammunition)
- 53rd Ordnance Company (Ammunition) (less Detachments)
- 56th Quartermaster Regiment, Company B
- 56th Quartermaster Regiment, Company B (Heavy Maintenance)
- 58th Armored Field Artillery Battalion
- 59th Signal Battalion
- 59th Armored Field Artillery Battalion
- 61st Armored Infantry Regiment
- 64th Armored Infantry Battalion
- 65th Armored Field Artillery Battalion
- 65th Field Artillery Regiment (Armored)
- 68th Armored Field Artillery Battalion
- 68th Field Artillery Regiment (Armored)
- 69th Armored Regiment
- 69th Armored Field Artillery Battalion
- 73rd Ordnance Company
- 73rd Ordnance Company (Detachment) (atchd)
- 75th Medical Battalion (Armored)
- 81st Armored Regiment
- 81st Armored Regimental Band
- 81st Armored Reconnaissance Battalion
- 81st Reconnaissance Battalion
- 85th Quartermaster Battalion, Company B (Light Maintenance)
- 85th Armored Reconnaissance Battalion
- 85th Reconnaissance Battalion
- 86th Engineer Battalion (Heavy Ponton)
- 87th Armored Field Artillery Battalion (105mm Howitzer, Truck-D) (AGF)
- 90th Quartermaster Company
- 90th Quartermaster Company (Railhead) (less Co HQ & 1 Platoon)
- 82nd AGF Band
- 86th Armored Reconnaissance Battalion

- 87th AGF Band
- 90th Quartermaster Company
- 91st Quartermaster Company (Railhead)
- 91st Armored Field Artillery Battalion
- 92nd Machine Records Unit (AGF)
- 92nd Machine Records Unit (Type B) (AGF)
- 93rd Armored Field Artillery Battalion
- 95th Armored Field Artillery Battalion
- 95th Field Artillery Battalion (Armored)
- 110th Cavalry Reconnaissance Troop, Mechanized
- 126th Separate Tank Company (Heavy) (AGF)
- 132nd AGF Band (AGF)
- 141st Armored Signal Company
- 141st Signal Company
- 145th Signal Company
- 145th Armored Signal Company
- 146th Armored Signal Company
- 151st Airborne Truck Company (AGF)
- 151st Antiaircraft Artillery Machine Gun Battery (Separate) (Airborne) w/atchd Medical
- 159th Ordnance Battalion, HQ & HQ Detachment watchd Medical Detachment (AGF)
- 159th Ordnance Battalion, HQ & HQ Detachment
- 447th Ordnance Heavy Auto Maintenance Company
- 448th Ordnance Heavy Auto Maintenance Company
- 449th Ordnance Heavy Auto Maintenance Company
- 918th Ordnance Heavy Auto Maintenance Company (Tank)
- 698th Ordnance Heavy Auto Maintenance Company (Tank)
- 699th Ordnance Heavy Auto Maintenance Company (Tank)
- 161st Airborne Engineer Battalin (less Company C)
- 189th Ordnance Depot Company (AGF)
- 196th AGF Band (Colored)
- 203rd Quartermaster Battalion, Company A (Gas Supply)
- 208th Quartermaster Gas Supply Battalion, HQ & HQ Detachment & Company A (AGF)
- 211th Ordnance Battalion, HQ & HQ Detachment
- 228th AGF Band
- 232nd Antiaircraft Artillery Searchlight Battalion (Type A) - 1 Platoon Battery C
- 350th Ordnance Battalion, HQ & HQ Detachment (AGF)
- 392nd Ordnance Medium Maintenance Company
- 485th Ordnance Evacuation Company (AGF)
- 486th Ordnance Evacuation Company (AGF)
- 553rd Ordnance Evacuation Company (AGF)
- 566th Ordnance Heavy Maintenance Company (Truck)
- 698th Ordnance Heavy Maintenance Company (Truck)
- 699th Ordnance Heavy Maintenance Company (Truck)
- 375th Engineer Battalion (Separate) (AGF)
- 375th Engineer Battalion (Separate) (Colored) (AGF)
- 376th Ordnance Heavy Auto Maintenance Company (AGF)
- 377th Engineer Battalion (Separate) (Colored) (AGF)

- 382nd Engineer Battalion (Separate) (AGF)
- 382nd Engineer Battalion (Separate) (Colored) (AGF)
- 387th Engineer Battalion (Separate) (AGF)
- 392nd Ordnance Medium Maintenance Company (AGF)
- 392nd Ordnance Medium Maintenance Company, Field A (AGF)
- 392nd Ordnance Heavy Maintenance Company, Tank (AGF)
- 395th Quartermaster Truck Company (Troop) (Colored) (AGF)
- 396th Quartermaster Truck Company (Troop) (Colored) (AGF)
- 397th Quartermaster Truck Company (Troop) (Colored) (AGF)
- 398th Quartermaster Truck Company (Troop) (Colored) (AGF)
- 400th Armored Field Artillery Battalion (AGF)
- 417th Ordnance Evacuation Company
- 425th Armored Field Artillery Battalion (less 1 Firing Battery)
- 425th Armored Field Artillery Battalion (less Battery A)
- 447th Ordnance Heavy Auto Maintenance Company (AGF)
- 448th Ordnance Heavy Auto Maintenance Company (AGF)
- 449th Ordnance Heavy Auto Maintenance Company (AGF)
- 453rd Antiaircraft Artillery Auto Weapons Battalion (Semi Mobile) (AGF)
- 470th Ordnance Evacuation Company (AGF)
- 476th Ordnance Evacuation Company (AGF)
- 485th Ordnance Evacuation Company (AGF)
- 486th Ordnance Evacuation Company (AGF)
- 487th Ordnance Evacuatin Company (AGF)
- 521st Quartermaster Group (Colored) (AGF)
- 521st Quartermaster Battalion, Mobile, HQ & HQ Detachment (Colored)
- 4021st Quartermaster Truck Company (Troop) (Colored)
- 4022nd Quartermaster Truck Company (Troop) (Colored)
- 4023rd Quartermaster Truck Company (Troop) (Colored)
- 4024th Quartermaster Truck Company (Troop) (Colored)
- 4029th Quartermaster Truck Company (Troop) (Colored)
- 4030th Quartermaster Truck Company (Troop) (Colored)
- 4031st Quartermaster Truck Company (Troop) (Colored)
- 4032nd Quartermaster Truck Company (Troop) (Colored)
- 105th Quartermaster Battalion, Mobile, HQ & HQ Detachment (Colored)
- 4025th Quartermaster Truck Company (Troop) (Colored)
- 4026th Quartermaster Truck Company (Troop) (Colored)
- 4027th Quartermaster Truck Company (Troop) (Colored)
- 4028th Quartermaster Truck Company (Troop) (Colored)
- 182nd Quartermaster Battalion, Mobile, HQ & HQ Detachment (Colored)
- 4029th Quartermaster Truck Company (Troop) (Colored)
- 4030th Quartermaster Truck Company (Troop) (Colored)
- 4031st Quartermaster Truck Company (Troop) (Colored)
- 4032nd Quartermaster Truck Company (Troop) (Colored)
- 521st Quartermaster Truck Regiment (Colored) (AGF)
- 527th Ordnance Heavy Maintenance Company (Tank) (AGF)
- 527th Armored Infantry Battalion (AGF)
- 552nd Quartermaster Railhead Company (AGF)

U.S. Army Installations - World War II:
A Research Reference

- 554th Ordnance Heavy Maintenance Company (Tank), Detachment (AGF)
- 565th Ordnance Heavy Maintenance Company (Truck) (AGF)
- 611th Tank Destroyer Battalion (Towed)
- 619th Ordnance Ammunition Company (AGF)
- 619th Ordnance Ammunition Company (Colored) (AGF)
- 651st Ordnance Ammunition Company (Colored) (AGF)
- 652nd Tank Destroyer Battalion
- 657th Engineer Topographic Battalion (A) - 1 Platoon
- 662nd Tank Destroyer Battalion (Towed)
- 671st Tank Destroyer Battalion (AGF)
- 698th Ordnance Heavy Maintenance Company (AGF)
- 698th Ordnance Heavy Maintenance Company (Tank) (AGF)
- 699th Ordnance Heavy Maintenance Company (AGF)
- 699th Ordnance Heavy Maintenance Company (Tank) (AGF)
- 701st Tank Destroyer Battalion (less 1 Reconnaissance Company)
- 717th Tank Battalion (AGF)
- 738th Tank Battalion, Medium (Special) (AGF)
- 739th Tank Battalion, Medium (Special) (AGF)
- 739th Tank Battalion, Medium (AGF)
- 739th Ordnance Heavy Auto Maintenance Company (AGF)
- 740th Tank Battalion (Medium) (AGF)
- 748th Tank Battalion (Medium) (Special) (AGF)
- 750th Tank Battalion (Medium) (AGF)
- 751st Tank Battalion (Medium)
- 764th Light Tank Battalion (AGF)
- 777th Infantry Regiment - Cannon Company (105mm Howitzer, Self-Propelled)
- 777th Tank Battalion (AGF)
- 779th Tank Battalion (AGF)
- 779th Tank Destroyer Battalion (less Company A)
- 779th Tank Battalion (less Cos A & B)
- 781st Tank Company (Separate) (AGF)
- 781st Tank Company (AGF)
- 781st Tank Battalion (Light) (AGF)
- 782nd Tank Battalion (AGF)
- 785th Tank Battalion
- 816th Tank Destroyer Battalion (Towed)
- 841st Ordnance Depot Company (AGF)
- 866th Ordnance Heavy Auto Maintenance Company (AGF)
- 891st Quartermaster Heavy Troop Transport Company (Colored)
- 892nd Quartermaster Troop Transport Company (Colored)
- 918th Ordnance Heavy Auto Maintenance Company (AGF)
- 969th Ordnance Heavy Auto Maintenance Company
- 3224th Ordnance Depot Company (AGF)
- 3234th Ordnance Depot Company (AGF)
- 3244th Ordnance Depot Company
- 3170th Ordnance Auto Maintenance Company (Provisional)
- 3171st Ordnance Auto Maintenance Company (Provisional)

4021st Quartermaster Truck Company (Troop) (Colored) (AGF)
4022nd Quartermaster Truck Company (Troop) (Colored) (AGF)
4029th Quartermaster Truck Company (Troop) (AGF)
4030th Quartermaster Truck Company (Troop) (AGF)
4031st Quartermaster Truck Company (Troop) (AGF)
4032nd Quartermaster Truck Company (Troop) (AGF)
Army Air Forces
Godman Field

Fort Thomas (III)
Newport, Kentucky

Fifth Corps Area Induction Station (Fort Thomas)
Army Service Forces
Provost Marshal Area #5 (I Act)
Reception Center
737th Military Police Battalion (Z of I) (ASF)
1516th Service Command Unit: Recruiting & Induction Substation (SOS)
1516th Service Command Unit: U.S. Army Recruiting Station (ASF)
1540th Service Command Unit: Station Complement (ASF)
1544th Corps Area Service Unit, Reception Center
1544th Service Command Unit: Reception Center (ASF)
3553rd Service Command Unit: Casual Company (ASF)
Army Air Forces
AAF Convalescent Hospital
Armed Forces Replacement Training Center
578th AAF Band
1076th AAF Base Unit (Convalescent Hospital)
1076th AAF Base Unit (Convalescent Hospital), WAC Squadron

Fort Hayes (I)
Columbus, Ohio

War Department Control
Inspector General's Department Officer Replacement Pool
Army Service Forces
Fifth Corps Area, Headquarters
    Inspector General's Department Officer Replacement Pool
    Finance Department Officer Replacement Pool (Coff)
    Judge Advocate General's Department Officer Replacement Pool (JAG)
    Sergeant Investigators, PMGO - Detachment #5 (PMGO)
    Fifth Service Command Records Depot
    Signal Corps Photographic Laboratory (ASF)
    Provost Marshal Area #3 (ASF)
    WAC Recruiting Service
1500th Corps Area Service Unit, Headquarters & Headquarters Company, Fifth Corps Area
1520th Corps Area Service Unit, Headquarters & Headquarters Company
1524th Corps Area Service Unit, Reception Center
1599th Corps Area Service Unit, (Detachment Corps of Intelligence Police Headquarters)

- Fifth Corps Area Engineers
- Fifth Corps Area Induction Station (Fort Hayes)
- Ohio Military District (Fifth Corps Area)
- Headquarters Fifth Service Command
- WAC Recruiting Service
- Fifth Service Command, HQ (SOS)
- Fifth Service Command Records Depot
- Army-Navy Procurement Boards, HQ (3 Traveling Boards) (ASF)
- Reclassification Center (ASF)
- Signal Corps Photographic Laboratory
- Provost Marshal Area #3
- Reception Center
- 5th Ordnance Service Company
- 8th Machine Readable Unit (Mobile) (AG) (Type Y)
- 18th Signal Service Company
- 36th Machine Records Unit (Mobile) (ASF)
- 36th Machine Records Unit (Mobile) (AG) (Type Z) (ASF)
- 44th Machine Records Unit (Mobile) (AG) (Type Y)
- 53rd Machine Readable Unit (Mobile) (AG) (Type Z) (ASF)
- 341st ASF Band
- 729th Military Police Battalion, Company D (ASF)
- 798th Military Police Battalion, Company A
- 1500th Service Command Unit: HQ & HQ Company (ASF)
- 1500th Service Command Unit: Headquarters Fifth Service Command
- 1507th Service Command Unit: Columbus Officer Procurement District
- 1508th Service Command Unit: Fifth Service Command ROTC Headquarters (ASF)
- 1510th Service Command Unit: Personnel Security Branch
- 1510th Service Command Unit: Investigation Division Corps of Military Police (ASF)
- 1511th Service Command Unit: Medical Examination & Induction Board, 5th SC, HQ
- 1513th Service Command Unit: Chief Maintenance Section, Automotive Branch (ASF)
- 1516th Service Command Unit: Recruiting & Induction Service, 5th SC, HQ & Substatn
- 1516th Service Command Unit: US Army Recruiting Station (ASF)
- 1520th Service Command Unit: US Army WAC Recruiting Station
- 1520th Service Command Unit: Station Complement
- 1520th Service Command Unit: Armed Forces Induction Station
- 1520th Service Command Unit: Signal Corps Repair Shop
- 1520th Service Command Unit: Station Complement
- 1520th Service Command Unit: Separation Point
- 1522nd Service Command Unit: Plant Protection, 5th SC, HQ & Columbus Dist. Office
- 1524th Service Command Unit: Reception Center
- 1533rd Service Command Unit: Signal Corps Repair Shop
- 1542nd Service Command Unit: Military Police Train Guard, 5th Service Command, HQ (AS
- 1558th Service Command Unit: Fifth Service Command Reclassification Center (ASF)
- 1558th Service Command Unit: Armed Forces Induction Station
- 1586th Service Command Unit: US Army Recruiting Station & AAF Examining Board
- 1594th Service Command Unit: Production Security District

1599th Service Command Unit: Counter Intelligence Corps, Fifth Service Command (ASF)
3551st Service Command Unit: Casual Company (ASF)

Camp Perry (I)
Lacarne, OH

Army Service Forces
1584th Corps Area Service Unit, Reception Center
Provost Marshal Area #4
Ordnance Unit Training Center Band (ASF)
Ordnance Officer Replacement Pool (ASF)
Ordnance Unit Training Center (ASF)
    1st Provisional Ordnance Unit Training Company
    2nd Provisional Ordnance Unit Training Company
    3rd Provisional Ordnance Unit Training Company
    4th Provisional Ordnance Unit Training Company
    5th Provisional Ordnance Unit Training Company
    6th Provisional Ordnance Unit Training Company
    7th Provisional Ordnance Unit Training Company
    8th Provisional Ordnance Unit Training Company
    9th Provisional Ordnance Unit Training Company
Ordnance Unit Training Center, HQ & HQ Detachment (ASF)
    5th Provisional Ordnance Unit Training Company
Reception Center (ASF)
15th Armored Division, Maintenance Battalion (AGF)
22nd Armored Division, Maintenance Battalion (ASF)
99th Italian Quartermaster Service Company (ASF)
100th Italian Engineer Base Depot Company (ASF)
100th Italian Quartermaster Service Company (ASF)
101st Italian Engineer Base Depot Company (ASF)
101st Italian Quartermaster Service Company (ASF)
147th Ordnance Motor Vehicle Assembly Company (Portable) (ASF)
199th Ordnance Depot Supply Company (SOS)
200th Ordnance Depot Supply Company (SOS)
253rd Ordnance Battalion, HQ & HQ Detachment & Medical Detachment (ASF)
254th Ordnance Battalion, HQ & HQ Detachment & Medical Detachment (ASF)
288th Ordnance Medium Maintenance Company (SOS)
289th Ordnance Medium Maintenance Company (SOS)
290th Ordnance Medium Maintenance Company (SOS)
291st Ordnance Medium Maintenance Company (SOS)
292nd Ordnance Medium Maintenance Company (SOS)
293rd Ordnance Medium Maintenance Company (SOS)
301st Ordnance Base Regiment, Company N (SOS)
304th Ordnance Base Regiment, Company O (ASF)
310th Italian Quartermaster Battalion, HQ & HQ Detachment (ASF)
394th Military Police Battalion (ZI) (Provisional)
408th Military Police Escort Guard Company (ASF)
412th Military Police Escort Guard Company (ASF)

Army Air Force

Fifth Service Command
Columbus Ohio

Army Air Force Bases

Indiana

Atterbury Army Air Field (III)

Columbus, IN
(Sub Base of Godman Field, KY)
(Inactive)
(Sub Base of George Field, IL)

Army Service Forces
3598th Service Command Unit: Service Group, Atterbury Army Air Field

Army Air Forces
Air Support Base (Observation) (serves Camp Atterbury)
AAF Weather Station
AAF Weather Station (Type C)
Finance Detachment
Medical Detachment
2nd Weather Squadron, Regional
2nd Weatehr Squadron, Regional, Detachment (Type E)
2nd Weatehr Squadron, Regional, Detachment (Type C)
2nd Army Airways Communications Squadron, Detachment
2nd Weather Squadron, Detachment
2nd Air Service Group, HQ & Base Services Squadron
2nd Air Service Group, 4th Air Engineering Squadron
2nd Air Service Group, 997th Air Materiel Squadron
53rd Station Complement Squadron (Special)
54th Station Complement Squadron (Special)
55th Station Complement Squadron (Special)
56th Station Complement Squadron (Special)
57th Station Complement Squadron (Special)
58th Station Complement Squadron (Special)
59th Station Complement Squadron (Special)
60th Station Complement Squadron (Special)
69th AAF Base Unit (2nd Weather Region), Section
85th AAF Base Unit, Section H (103rd AACS Squadron), Detachment
118th AAF Base Unit (Bomb (Medium))
304th Base Headquarters & Air Base Squadron
431st Sub-Depot
590th Air Materiel Squadron - Stock Control & Warehouse Section Det (Colored)
602nd Air Engineering Squadron (less Shop, Vehicle Detachment)
618th Bomb Squadron, Mediuim (Colored)
619th Bomb Squadron, Mediuim (Colored)

733rd AAF Base Unit (103rd AACS Squadron), Detachment
805th AAF Base Unit (CCTS, Troop Carrier)
853rd Signal Service Company, Aviation, Detachment
903rd Quartermaster Service Company, Aviation, Detachment #39
1035th Guard Squadron
2065th Ordnance Company, Aviation (Service), Detachment

Baer Field (III)
Fort Wayne, IN
Army Service Forces
3565th Service Command Unit: Service Group, Baer Field
Army Air Forces
I Troop Carrier Processing Unit
AAF Staging Area
AAF Staging Area for Air Departure
AAF Weather Station (Type A)
Finance Detachment
Medical Detachment
Veterinary Detachment
WAC Detachment
Air Freight Terminal, AFATC
Central Assembly Station
Third Echelon Repair Shop
IX Troop Carrier Command (Special), HQ & HQ Squadron
1st Chemical Storage Company (Aviation) (less Decon Detachment)
1st Reception & Final Phase Unit, Troop Carrier Command
1st Air Cargo Resupply Detachment
2nd Army Airways Communications Squadron, Detachment
2nd Weather Squadron, Detachment
2nd Weather Squadron, Regional, Detachment (Type A)
3rd Air Cargo Resupply Detachment
5th Sub-Depot
8th Airdrome Squadron
9th Airdrome Squadron
10th Service Group, HQ & HQ Squadron
13th Service Squadron
337th Service Squadron
18th Airdrome Squadron
19th Airdrome Squadron
23rd Statistical Control Unit, Overseas (Special)
24th Signal Platoon (Air Borne)
30th Quartermaster Regiment, Company B (Truck)
31st Pursuit Group (Interceptor), HQ & HQ Squadron
39th Pursuit Squadron
40th Pursuit Squadron
41st Pursuit Squadron
45th Base Headquarters & Air Base Squadron
46th Air Base Group, HQ & HQ Squadron

- 45th Air Bomb Squadron
- 61st Materiel Squadron
- 52nd Troop Carrier Wing, HQ & HQ Squadron
- 64th Aviation Squadron (Separate)
- 64th Aviation Squadron (Colored)
- 69th AAF Base Unit (2nd Weather Region), Section
- 71st AAF Band
- 85th AAF Base Unit, Section H (103rd AACS Squadron), Detachment
- 86th Transportation Squadron (Cargo & Mail)
- 87th Transportation Squadron (Cargo & Mail)
- 88th Quartermaster Regiment, 2nd Platoon, Company A (Light Maintenance) (less Det)
- 98th Service Group, HQ & HQ Squadron
- 99th Signal Company, Service Group
- 313th Troop Carrier Group, Headquarters
- 29th Troop Carrier Squadron
- 47th Troop Carrier Squadron
- 48th Troop Carrier Squadron
- 49th Troop Carrier Squadron
- 321st Transportation Squadron (Cargo & Mail)
- 329th Transportation Squadron (Cargo & Mail)
- 329th Airdrome Squadron (Special)
- 330th Airdrome Squadron (Special)
- 331st Airdrome Squadron (Special)
- 332nd Airdrome Squadron (Special)
- 349th Troop Carrier Group, Headquarters
- 23rd Troop Carrier Squadron
- 312th Troop Carrier Squadron
- 313th Troop Carrier Squadron
- 314th Troop Carrier Squadron
- 434th Troop Carrier Group, Headquarters
- 71st Troop Carrier Squadron
- 72nd Troop Carrier Squadron
- 73rd Troop Carrier Squadron
- 74th Troop Carrier Squadron
- 435th Troop Carrier Group, Headquarters
- 75th Troop Carrier Squadron
- 76th Troop Carrier Squadron
- 77th Troop Carrier Squadron
- 78th Troop Carrier Squadron
- 436th Troop Carrier Group, Headquarters
- 79th Troop Carrier Squadron
- 80th Troop Carrier Squadron
- 81st Troop Carrier Squadron
- 82nd Troop Carrier Squadron
- 437th Troop Carrier Group, Headquarters
- 83rd Troop Carrier Squadron
- 84th Troop Carrier Squadron
- 85th Troop Carrier Squadron
- 86th Troop Carrier Squadron

- 438th Troop Carrier Group, Headquarters
- 87th Troop Carrier Squadron
- 88th Troop Carrier Squadron
- 89th Troop Carrier Squadron
- 90th Troop Carrier Squadron
- 439th Troop Carrier Group, Headquarters
- 91st Troop Carrier Squadron
- 92nd Troop Carrier Squadron
- 93rd Troop Carrier Squadron
- 94th Troop Carrier Squadron
- 440th Troop Carrier Group, Headquarters
- 95th Troop Carrier Squadron
- 96th Troop Carrier Squadron
- 97th Troop Carrier Squadron
- 98th Troop Carrier Squadron
- 441st Troop Carrier Group, Headquarters
- 99th Troop Carrier Squadron
- 100th Troop Carrier Squadron
- 301st Troop Carrier Squadron
- 302nd Troop Carrier Squadron
- 442nd Troop Carrier Group, Headquarters
- 303rd Troop Carrier Squadron
- 304th Troop Carrier Squadron
- 305th Troop Carrier Squadron
- 306th Troop Carrier Squadron
- 443rd Troop Carrier Group, Headquarters
- 309th Troop Carrier Squadron
- 310th Troop Carrier Squadron
- 452nd Air Service Group, HQ & Base Services Squadron
- 828th Air Engineering Squadron, Air Service Group
- 652nd Air Materiel Squadron, Air Service Group
- 453rd Air Service Group, HQ & Base Services Squadron
- 871st Air Engineering Squadron, Air Service Group
- 695th Air Materiel Squadron, Air Service Group
- 455th Air Service Group, HQ & Base Services Squadron
- 673rd Air Engineering Squadron, Air Service Group
- 697th Air Materiel Squadron, Air Service Group
- 460th Air Service Group, HQ & Base Services Squadron
- 878th Air Engineering Squadron, Air Service Group
- 702nd Air Materiel Squadron (Air Service Group)
- 461st Air Service Group, HQ & Base Services Squadron
- 879th Air Engineering Squadron, Air Service Group
- 703rd Air Materiel Squadron (Air Service Group)
- 477th Air Service Group, HQ & Base Services Squadron
- 895th Air Engineering Squadron, Air Service Group
- 719th Air Materiel Squadron, Air Service Group
- 502nd AAF Base Unit (HQ, AFATC Field Authorization - Operating Location #35
- Staging Area Office (AFATC Lisiaon Office)
- Office of Route Control

- 553rd AAF Ferrying Group (3rd Ferrying Group) – Operating Location #17
- Military Air Transport Station
- 567th AAF Band
- 571st AAF Band
- 679th Ordnance Company (Aviation) (Pursuit)
- 709th Ordnance Company (Aviation) (Air Borne)
- 733rd AAF Base Unit (103rd AACS Squadron), Detachment
- 753rd WAC Post Headquarters Company, AAF
- 806th AAF Base Unit (Processing Outbound, Troop Carrier)
- 806th AAF Base Unit (Processing Outbound, Troop Carrier), WAC Squadron
- 849th Guard Squadron
- 859th Signal Service Company (Aviation), Detachment
- 867th Signal Service Company, Aviation, Detachment
- 905th Quartermaster Service Company, Aviation, Detachment #3
- 915th Quartermaster Service Company, Aviation, Detachment
- 1007th Ordnance Company, Aviation, Airborne
- 1063rd Quartermaster Company, Service Group, Aviation
- 1506th Service Command Unit: Sub-School for Bakers & Cooks #6 (SOS)
- 1802nd Ordnance Medium Maintenance Company (Aviation) (Q)
- 1829th Ordnance Medium Maintenance Company (Aviation) (Q)
- 1830th Ordnance Medium Maintenance Company (Aviation) (Q)
- 2047th Quartermaster Truck Company, Aviation
- 2048th Quartermaster Truck Company, Aviation
- 2054th Ordnance Service Company, Aviation, Detachment
- 2076th Quartermaster Truck Company, Aviation
- 2077th Quartermaster Truck Company, Aviation

Bendix Field (III)
South Bend, IN (South Bend Municipal Airport)
(Temporarily Inactive)
(Surplus)
Army Air Forces
Air Transport Base (Ferry)
1st Ferrying Service Detachment (S)
69th AAF Base Unit (2nd Weather Region), Section
570th AAF Base Unit (1st Ferrying Service Station)
570th AAF Base Unit (1st Ferrying Service Station), WAC Squadron

Columbus Municipal Airport
Columbus, IN *(Converted to Atterbury Army Air Base)*
Army Air Force
Air Support Base (Observation)

Emison Auxiliary Field #1 (III)
Vincennes, IN
(Auxiliary of George Field, IL)

Evansville Municipal Airport (III)

Evansville, IN
Army Air Forces
Republic-Evansville AAF Mofication Center #6
Air Freight Terminal, AFATC
2nd Weather Squadron, Detachment
69th AAF Base Unit (2nd Weather Region), Section
553rd AAF Base Unit (3rd Ferrying Group) - Operation Location #15
Military Air Transport Station
Modification and Factory Control Offices
Ferrying Control Office
554th AAF Base Unit (4th Ferry Group) - Operating Location #2
Modification and Factory Control Offices
Factory Control Office
554th AAF Base Unit (HQ, ATC Field Authorization) - Operating Location

Freeman Field (III)
Seymour, IN
(Inactive)
(Sub Base of Godman Field, KY)
Army Service Forces
3569th Service Command Unit: Post Engineers (ASF)
3569th Service Command Unit: Service Group, Freeman Field (ASF)
Army Air Forces
AAF Advanced Flying School (Twin Engine)
AAF Weather Station (Type B)
AAF Weather Station (Type A)
AAF Pilot School (Advanced 2-Engine)
Aviation Cadet Training Group
Finance Detachment
Medical Detachment
Veterinary Detachment
WAC Detachment
Third Echelon Repair Shop
2nd Army Airways Communications Squadron, Detachment
2nd Weather Squadron - Regional, Detachment
2nd Weather Squadron, Regional, Detachment (Type A)
2nd Weather Squadron, Detachment
35th Two-Engine Flying Training Group, HQ & HQ Squadron
36th Two-Engine Flying Training Group, HQ & HQ Squadron
69th AAF Base Unit (2nd Weather Region), Section
85th AAF Base Unit, Section H (103rd AACS Squadron), Detachment
118th AAF Base Unit (Bomb (Medium))
320th Aviation Squadron
320th Aviation Squadron (Colored)
366th Sub-Depot
387th Air Service Group (Colored)
Headquarters & Base Services Squadron (Colored)
590th Air Materiel Squadron (Colored)
602nd Air Materiel Squadron (Colored)

405th AAF Band
406th AAF Band
447th Base Headquarters & Air Base Squadron
466th Two-Engine Flying Training Squadron
467th Two-Engine Flying Training Squadron
477th Bombardment Group, Medium (Colored)
　　477th Bombardment Group, HQ (Colored)
　　616th Bombardment Squadron, Medium (Colored)
　　617th Bombardment Squadron, Medium (Colored)
　　618th Bombardment Squadron, Medium (Colored)
　　619th Bombardment Squadron, Medium (Colored)
705th AAF Band
733rd AAF Base Unit (103rd AACS Squadron), Detachment
742nd WAC Post Headquarters Company, AAF
766th AAF Band (Colored)
856th Signal Service Company, Aviation, Detachment
907th Quartermaster Service Company, Aviation, Detachment
1079th Two-Engine Flying Training Squadron
1080th Two-Engine Flying Training Squadron
1087th Guard Squadron
2062nd Ordnance Company, Aviation (Service), Detachment
2139th AAF Base Unit (Pilot School, Advanced 2-Engine)
2139th AAF Base Unit (Pilot School, Advanced 2-Engine), WAC Squadron
4120th AAF Base Unit (Air Base)

Grammer Auxiliary Field #3
Grammer, IN
(Auxiliary of Freeman Field, IN)
(Surplus)

Indiana State Fairgrounds
Indianapolis, IN
Army Service Forces
3578th Service Command Unit: Post Engineers, AAF Storage Depot (ASF)
3578th Service Command Unit: Service Group, 836th AAF Specialized Depot (ASF)
Army Air Forces
AAF Armament School
AAF Technical School (Armament #3)
AAF Storage Depot
AAF Depot Training Station
Medical Detachment
4th AAF Storage Depot
836th AAF Specialized Depot (less Detachments A, B & C) (AAF)
836th AAF Specialized Depot (PZ 5)
4836th AAF Base Unit (Specialized Depot) (PZ 5)
4836th AAF Base Unit (Specialized Depot) (PZ 5), WAC Squadron

Indianapolis Municipal Airport (Ill)

Indianapolis, IN
    AAF

Madison Army Air Field
Jefferson Proving Ground, IN
    Army Air Force
    Jefferson Proving Ground (see Jefferson Proving Ground (Army Service Forces))
    AAF Weather Station (Type D)
    3rd Proving Ground Detachment

Milport Auxiliary Field #4 (III)
Millport, IN
    (Auxiliary of Freeman Field, IN)
    (Auxiliary of Godman Field, KY)

Milroy Intermediate Field (III)
Milroy, IN
    (Auxiliary of Bowman Field, KY)

Richmond Municipal Airport (III)
Richmond, IN
    AAF

St. Anne Auxiliary Field #2 (III)
North Vernon, IN
    (Auxiliary of Freeman Field, IN)
    (Auxiliary of Godman Field, KY)

Schoen Field (III)
Fort Benjamin Harrison, IN
    Army Air Force
    Landing Field (Fort Benjamin Harrison)

Seymour Advanced Flying School *(Converted to Freeman Field)*
Seymour, IN
    Army Air Force
    Medical Detachment
    Veterinary Detachment
    320th Aviation Squadron (Separate)
    366th Sub Depot
    447th Base Headquarters & Air Base Squadron
    856th Signal Service Company, Aviation, Detachment
    907th Quartermaster Service Company, Aviation, Detachment
    1077th Two-Engine Flying Training Squadron
    1078th Two-Engine Flying Training Squadron
    1079th Two-Engine Flying Training Squadron
    1080th Two-Engine Flying Training Squadron

1087th Guard Squadron

South Bend Municipal Airport
South Bend, IN   *(Converted to Bendix Field, IN)*

Army Air Force
Air Transport Base (Ferry)
1st Ferrying Service Detachment (S)

Stout Field (III)
Indianapolis, IN

Army Service Forces
3567th Service Command Unit: Post Engineers (ASF)
3567th Service Command Unit: Service Group, Stout Field (ASF)

Army Air Forces
I Troop Carrier Command, HQ (or) (HQ, I Troop Carrier Command)
I Troop Carrier Command, HQ & HQ Squadron
Base Weather Station
AAF Weather Station (Type A)
Finance Detachment
Medical Detachment
Veterinary Detachment
WAC Detachment
Air Freight Terminal, AFATC
IX Troop Carrier Command, HQ & HQ Squadron
2nd Weather Squadron, Detachment
2nd Weather Squadron, Regional, Detachment
2nd Weather Squadron, Regional, Detachment (Type A)
17th Statistical Control Unit
17th Statistical Control Unit (Type C)
39th Air Freight Wing, Detachment
69th AAF Base Unit (2nd Weather Region), Section
362nd Base Headquarters & Air Base Squadron
367th AAF Band
389th Signal Company, Aviation (less Radio Intelligence Platoon)
438th Aviation Squadron (Colored)
553rd AAF Base Unit (3rd Ferrying Group) - Operating Location #12
Air Freight and Passenger Handling
Military Air Transport Station
667th AAF Band
704th AAF Band
712th WAC Post Headquarters Company, AAF
733rd AAF Base Unit (103rd AACS Squadron), Detachment
800th AAF Base Unit (Administrative Unit, I Troop Carrier Command)
800th AAF Base Unit (Administrative Unit, ITCC), WAC Squadron
814th AAF Base Unit (Glider Testing & Ferrying Troop Carrier)
814th AAF Base Unit (Glider Testing & Ferrying Troop Carrier), WAC Squadron
859th Signal Service Company, Aviation, Detachment
867th Signal Service Company, Aviation (less Detachments) (less personnel)

867th Signal Service Company, Aviation (less Detachments)
915th Quartermaster Service Company, Aviation
915th Quartermaster Service Company, Aviation (less Dets) (less personnel)
915th Quartermaster Service Company, Aviation (less Dets)
930th Guard Squadron
3521st AAF Base Unit (Factory School) (Allison Division, GMC)
2068th Ordnance Company, Aviation (Service) (less personnel)
2068th Ordnance Company, Aviation (Service)

Vincennes Municipal Airport (III)
Vincennes, IN
AAF

Walesboro Auxiliary Field #1 (III)
Walesboro, IN
(Auxiliary of Freeman Field, IN)
(Auxiliary of George Field, IL)

Zenas Auxiliary Field #5
Zenas, IN
(Auxiliary of Freeman Field, IN)
(Surplus)

Kentucky

Anchorage Field (III)
Anchorage, KY
(Auxiliary of Bowman Field, KY)

Bowling Green Municipal Airport (III)
Bowling Green, KY
AAF
11th Tactical Reconnaissance Squadron
28th Tactical Reconnaissance Squadron
88th Airdrome Squadron
559th Antiaircraft Artillery Auto Weapons Battalion (Mobile)

Bowman Field (III)
Louisville, KY
Army Service Forces
1567th Service Command Unit: School for Bakers & Cooks (ASF)
3566th Service Command Unit: Service Group, Bowman Field (ASF)
3566th Service Command Unit: Service Group: Post Engineers (ASF)
Army Air Forces
AAF School of Air Evacuation
Glider Crew Training Center

- Finance Detachment
- Medical Detachment
- Veterinary Detachment
- WAC Detachment
- AAF Weather Station (Type A)
- AAF Convalescent Hospital
- Air Freight Terminal, AFATC
- Third Echelon Repair Shop
- Troop Carrier Command Glider Pilot Combat Training Unit
- 1st Provisional Group
- 1st Combat Cargo Group, Headquarters
- 1st Combat Cargo Squadron
- 2nd Combat Cargo Squadron
- 3rd Combat Cargo Squadron
- 4th Combat Cargo Squadron
- 2nd Army Airways Communications Squadron, Detachment
- 2nd Weather Squadron, Detachment (Type A)
- 2nd Weather Squadron, Regional, Detachment (Type A)
- 4th Combat Cargo Group, Headquarters
- 13th Combat Cargo Squadron
- 14th Combat Cargo Squadron
- 15th Combat Cargo Squadron
- 16th Combat Cargo Squadron
- 5th Air Support Command, HQ & HQ Squadron
- 8th Reconnaissance Squadron (Light)
- 16th Sub-Depot
- 16th Bombardment Wing, Headquarters
- 18th Airdrome Squadron
- 21st Signal Platoon (Airborne)
- 19th Airdrome Squadron
- 26th AAF Base Unit (AAF School of Air Evacuation)
- 27th Base Headquarters & Air Base Squadron
- 28th Air Base Group, HQ & HQ Squadron
- 27th Air Force Squadron
- 39th Materiel Squadron
- 34th Troop Carrier Squadron - Glider Echelon (Provisional)
- 38th Troop Carrier Squadron - Glider Echelon (Provisional)
- 38th Troop Carrier Squadron
- 43rd Troop Carrier Squadron - Glider Echelon (Provisional)
- 43rd Aviation Squadron (Colored)
- 46th Bombardment Group (Light), HQ & HQ Squadron
- 50th Bombardment Squadron
- 51st Bombardment Squadron
- 53rd Bombardment Squadron
- 67th AAF Band
- 69th AAF Base Unit (2nd Weather Region), Section
- 87th Bombardment Squadron (Light) (46th Bombardment Group)
- 88th Quartermaster Battalion, Detachment-2nd Platoon, Company A (Light Maintenance)
- 93rd Quartermaster Battalion, Company D (Light Maintenance)

- 93rd Quartermaster Battalion, 2nd Platoon, Company C (Light Maintenance) (less Det)
- 286th Quartermaster Company (Airborne)
- 322nd Signal Company (Air Wing)
- 344th Airdrome Squadron (Special)
- 345th Airdrome Squadron (Special)
- 346th Airdrome Squadron (Special)
- 347th Airdrome Squadron (Special)
- 348th Airdrome Squadron (Special)
- 349th Air Evacuation Group, HQ & HQ Squadron
- 622nd Air Evacuation Squadron (Light)
- 803rd Medical Squadron Air Evacuation Transport
- 804th Medical Squadron Air Evacuation Transport
- 805th Medical Squadron Air Evacuation Transport
- 806th Medical Squadron Air Evacuation Transport
- 807th Medical Squadron Air Evacuation Transport
- 808th Medical Squadron Air Evacuation Transport
- 350th Airdrome Squadron (Special)
- 351st Airdrome Squadron (Special)
- 375th Troop Carrier Group, Headquarters
- 55th Troop Carrier Squadron
- 56th Troop Carrier Squadron
- 57th Troop Carrier Squadron
- 58th Troop Carrier Squadron
- 422nd Signal Company (Aviation)
- 433rd Ordnance Company (Aviation) (Bombardment)
- 443rd Ordnance Company (Aviation) (Bombardment)
- 553rd AAF Base Unit (3rd Ferrying Group) - Operating Location #1
- Modification Control Office
- Military Air Transport Station
- 567th AAF Band
- 706th Ordnance Company (Aviation) (Airborne)
- 707th Ordnance Company (Aviation) (Airborne)
- 717th WAC Post Headquarters Company, AAF
- 745th AAF Band
- 805th Medical Air Evacuation Transport Squadron (less Detachment & 1 Flight)
- 808th AAF Base Unit (Operational Training Troop Carrier)
- 808th AAF Base Unit (Operational Training Troop Carrier), WAC Section
- 808th Medical Air Evacuation Transport Squadron (less Detachment & 1 Flight)
- 809th Medical Air Evacuation Transport Squadron (less Detachment & 1 Flight)
- 811th Medical Air Evacuation Transport Squadron (less Detachment & 1 Flight)
- 812th Medical Air Evacuation Transport Squadron (less Detachment & 1 Flight)
- 815th Medical Air Evacuation Transport Squadron
- 816th Medical Air Evacuation Transport Squadron
- 817th Medical Air Evacuation Transport Squadron
- 818th Medical Air Evacuation Transport Squadron
- 819th Medical Air Evacuation Transport Squadron
- 820th Medical Air Evacuation Transport Squadron

- 821st Medical Air Evacuation Transport Squadron
- 822nd Medical Air Evacuation Transport Squadron
- 822nd Medical Air Evacuation Transport Squadron, Squadron HQ & Supply Section
- 823rd Medical Air Evacuation Transport Squadron
- 823rd Medical Air Evacuation Transport Squadron (less Flights A, B, C & D)
- 824th Medical Air Evacuation Transport Squadron
- 824th Medical Air Evacuation Transport Squadron (less Evacuation Flights A & B)
- 825th Medical Air Evacuation Transport Squadron
- 825th Medical Air Evacuation Transport Squadron, HQ
- 826th Medical Air Evacuation Transport Squadron (less HQ)
- 826th Medical Air Evacuation Transport Squadron
- 827th Medical Air Evacuation Transport Squadron, HQ
- 827th Medical Air Evacuation Transport Squadron (less HQ)
- 828th Medical Air Evacuation Transport Squadron
- 828th Medical Air Evacuation Transport Squadron, HQ
- 829th Medical Air Evacuation Transport Squadron
- 829th Medical Air Evacuation Transport Squadron, HQ
- 830th Medical Air Evacuation Transport Squadron (less Flights A & B)
- 830th Medical Air Evacuation Transport Squadron, HQ (Flights 31 thru 39)
- 867th Signal Service Company, Aviation, Detachment
- 915th Quartermaster Service Company, Aviation, Detachment
- 926th Quartermaster Platoon Transport Air Base
- 952nd Guard Squadron
- 1077th AAF Base Unit (Convalescent Hospital)
- 1077th AAF Base Unit (Convalescent Hospital), WAC Squadron
- 1506th Service Command Unit: Sub-School for Bakers & Cooks #7 (SOS)
- 1567th Service Command Unit: School for Bakers & Cooks (ASF)
- 2068th Ordnance Company, Aviation (Service), Detachment

Campbell Army Air Field (III)
Clarksville, TN (HQ in Kentucky)
(Sub-Base of Godman Field, KY)
(Sub-Base of William Norther Field, TN)
(Auxiliary of Smyrna Army Air Field, TN)

Army Service Forces
- 3599th Service Command Unit: Service Group, Campbell Army Air Field (ASF)
- 3599th Service Command Unit: Regional Post Engineer, Paducah Area (ASF)

Army Air Forces
- Air Support Base (Observation) (serves Camp Campbell, KY)
- AAF Weather Station (Type D)
- AAF Weather Station (Type C)
- 2nd Army Airways Communications Squadron, Detachment
- 2nd Weather Squadron, Regional
- 2nd Weather Squadron, Regional, Detachment (Type C)
- 2nd Weather Squadron, Detachment
- 3rd Photo Lab Section
- 6th Photo Lab Section
- 31st AAA Group, HQ & HQ Battery

31st Service Group, HQ & HQ Squadron
56th Service Squadron
305th Service Squadron
36th Photo Mapping Squadron
66th Reconnaissance Group, Headquarters
19th Liaison Squadron
97th Reconnaissance Squadron (Fighter)
106th Reconnaissance Squadron (Bomber)
118th Reconnaissance Squadron (Fighter)
69th AAF Base Unit (2nd Weather Region), Section
73rd Tactical Reconnaissance Group, Headquarters
15th Tactical Reconnaissance Squadron
152nd Tactical Reconnaissance Squadron
74th Tactical Reconnaissance Group, Headquarters
11th Tactical Reconnaissance Squadron
13th Tactical Reconnaissance Squadron
22nd Tactical Reconnaissance Squadron
85th AAF Base Unit, Section H (103rd AACS Squadron), Detachment
86th Service Squadron
99th Base Headquarters & Air Base Squadron, Detachment
343rd Base Headquarters & Air Base Squadron, Detachment
510th AAA Gun Battalion (Semi Mobile)
541st AAA Auto Weapons Battalion (Semi Mobile)
733rd AAF Base Unit (103rd AACS Squadron), Detachment
787th AAA Auto Weapons Battalion (Semi Mobile)
853rd Signal Service Company, Aviation, Detachment
1009th Signal Company, Service Group, Detachment
1033rd Guard Squadron – Security Detachment #1
1039th Signal Company, Service Group
1098th Quartermaster Company, Service Group, Aviation, Detachment
1112th Quartermaster Company, Service Group, Aviation
1731st Ordnance Supply & Maintenance Company, Aviation
1732nd Ordnance Supply & Maintenance Company, Aviation
1760th Ordnance Supply & Maintenance Company, Aviation
1916th Quartermaster Truck Company, Aviation
2003rd Quartermaster Truck Company, Aviation (Colored)
2007th Quartermaster Truck Company, Aviation

Godman Field (III)
Fort Knox, KY
(Sub Base of Freeman Field)

Army Service Forces
3549th Service Command Unit: Service Group, Godman Field (ASF)

Army Air Forces
Air Support Base (Observation) (serves Fort Knox, KY)
AAF Weather Station (Type B)
AAF Weather Station (Type A)
Finance Detachment
Medical Detachment

- Medical Detachment, Dispensary
- Veterinary Detachment
- Veterinary Detachment, Dispensary
- WAC Detachment
- Third Echelon Repair Shop
- 2nd Army Airways Communications Squadron, Detachment
- 2nd Weather Squadron, Detachment
- 2nd Weather Squadron, Regional, Detachment
- 2nd Weather Squadron, Regional, Detachment (Type A)
- 2nd Observation Squadron, Detachment
- 31st Quartermaster Regiment, Company B (Truck) (Colored)
- 34th Quartermaster Regiment, Companies H & K, (Truck)
- 34th Quartermaster Regiment, Company K (Truck)
- 69th AAF Base Unit (2nd Weather Region), Section
- 73rd Reconnaissance Group, Headquarters
- 14th Liaison Squadron
- 15th Reconnaissance Squadron (Fighter)
- 28th Reconnaissance Squadron (Righter)
- 91st Reconnaissance Squadron (Bomber)
- 73rd Observation Group, Headquarters
- 15th Observation Squadron (less Air Echelon)
- 28th Observation Squadron
- 91st Observation Squadron
- 85th AAF Base Unit, Section H (103rd AACS Squadron), Detachment
- 91st Tactical Reconnaissance Squadron
- 99th Base Headquarters & Air Base Squadron
- 99th Base Headquarters & Air Base Squadron (less Detachment)
- 115th AAF Base Unit (Bomb (Medium))
- 115th AAF Base Unit (Bomb (Medium)) (less Det Sec C & Det Sec G)
- 118th AAF Base Unit (Bomb (Medium))
- 118th AAF Base Unit (Bomb (Medium) & Fighter)
- 248th Quartermaster Company (Airborne)
- 249th Quartermaster Company (Airborne)
- 387th Air Service Group HQ & Base Services Squadron (Colored)
- 590th Air Materiel Squadron (less Stock Contl & Whse Sec Det) (Colored)
- 602nd Air Engineering Squadron – Shop Vehicle Detachment (Colored)
- 391st Bombardment Group (Medium), Headquarters
- 573rd Bombardment Squadron
- 574th Bombardment Squadron
- 575th Bombardment Squadron
- 412th Sub-Depot
- 477th Bomb Group (Medium), HQ (Colored)
- 616th Bomb Squadron, Medium (Colored)
- 617th Bomb Squadron, Medium (Colored)
- 619th Bomb Squadron, Medium (Colored)
- 477th Composite Group (Colored), Headquarters
- 99th Fighter Squadron (SE) (Colored)
- 617th Bombardment Squadron, Medium (Colored)
- 618th Bombardment Squadron, Medium (Colored)

766th AAF Band (Colored)
827th Chemical Company, Air Operations (M & H)
847th WAC Post Headquarters Company, AAF
853rd Signal Service Company, Aviation, Detachment
903rd Quartermaster Service Company, Aviation, Detachment #26
1008th Guard Squadron
2065th Ordnance Company, Aviation (Service), Detachment

Kenton County Airport
Covington, KY
    Air Transport Base

Lexington Municipal Airport (III)
Lexington, KY
    AAF
    (Auxilliary of Bowman Field, KY)

Louisville Municipal Airport #2 (III)
Louisville, KY (Standiford Field)
    Army Air Forces
    Consolidated Vultee Modificaiton Center #9
    69th AAF Base Unit (2nd Weather Region), Section
    86th AAF Technical Training Detachment (Consolidated Vultee Modification Center #9)
    553rd AAF Base Unit (3rd Ferrying Group) - Operating Location #1
        Modification and Factory Control Office
    553rd AAF Base Unit (3rd Ferrying Group) - Operating Location #18
        Modification and Factory Control Office
        Ferrying Control Office

Paducah Municipal Airport (III)
Paducah, KY
    (Stand-by Status)
    (Inactive)
    Army Air Force
    Air Transport Command

Standiford Field
Louisville, KY
    (See Louisville Municipal Airport #2)
    Army Air Force
    Vultee-Louisville AAF Modification Center #9
        Training Detachment AAFTTC

Sturgis Army Air Field (III)
Sturgis, KY (Morganfield Municipal Airport)
    (Sub Bse of Godman Field, KY)
    (Sub Base of Key Field, MS)

AAF Weathr Station (Type D)
2nd Army Airways Communications Squadron, Detachment
2nd Weather Squadron, Detachment
2nd Weather Squadron, Regional, Detachment (Type D)
69th AAF Base Unit (2nd Weather Region), Section
85th AAF Base Unit, Section H (103rd AACS Squadron), Detachment
703rd AAF Base Unit (Glider Test)
733rd AAF Base Unit (103rd AACS Squadron), Detachment
4143rd AAF Base Unit (Glider Test)
1072nd Guard Squadron, Detachment

Dayton Municipal Airport (III)
Dayton, OH
AAF
Northwest-Vandalia Modification Center #11
702nd AAF Base Unit (Accelerated Service Text)
1072nd Guard Squadron, Detachment

Dayton Army Air Field (III)
Dayton, OH
(Sub Base of Wright Field, OH)
Army Air Forces
Northwest-Vandalia Modification Center #11
Field Office of Materiel Division
502nd AAF Base Unit (HQ, AFATC Field Authorization) - Operating Location #8
Regional Air Priorities Control
ATC Liaison Office
AFATC Liaison Office
Priority Control Office
702nd AAF Base Unit (Accelerated Service Test)
1072nd Guard Squadron, Detachment
4142nd AAF Base Unit (Accelerated Service Test)

Lockbourne Army Air Base (III)
Columbus, OH
Army Service Forces
3568th Service Command Unit: Post Engineers (ASF)
3568th Service Group, Lockbourne Army Air Base (ASF)
Army Air Forces
AAF Combat Crew School
AAF Glider School
Base Weather Station
Glider Pilot Replacement Pool
AAF Instructors School (Pilot 4-Engine)
AAF Pilot School (Specialized 4-Engine)
AAF Weather Station (Type B)
AAF Weather Station (Type A)
Finance Detachment

AAF Weatehr Station (Type D)
2nd Army Airways Communications Squadron, Detachment
2nd Weather Squadron, Detachment
2nd Weather Squadron, Regional, Detachment (Type D)
69th AAF Base Unit (2nd Weather Region), Section
85th AAF Base Unit, Section H (103rd AACS Squadron), Detachment
703rd AAF Base Unit (Glider Test)
733rd AAF Base Unit (103rd AACS Squadron), Detachment
4143rd AAF Base Unit (Glider Test)
1072nd Guard Squadron, Detachment

Dayton Municipal Airport (III)
Dayton, OH
AAF
Northwest-Vandalia Modification Center #11
702nd AAF Base Unit (Accelerated Service Text)
1072nd Guard Squadron, Detachment

Dayton Army Air Field (III)
Dayton, OH
(Sub Base of Wright Field, OH)
Army Air Forces
Northwest-Vandalia Modification Center #11
Field Office of Materiel Division
502nd AAF Base Unit (HQ, AFATC Field Authorization) - Operating Location #8
  Regional Air Priorities Control
  ATC Liaison Office
  AFATC Liaison Office
  Priority Control Office
702nd AAF Base Unit (Accelerated Service Test)
1072nd Guard Squadron, Detachment
4142nd AAF Base Unit (Accelerated Service Test)

Lockbourne Army Air Base (III)
Columbus, OH
Army Service Forces
3568th Service Command Unit: Post Engineers (ASF)
3568th Service Group, Lockbourne Army Air Base (ASF)
Army Air Forces
AAF Combat Crew School
AAF Glider School
Base Weather Station
Glider Pilot Replacement Pool
AAF Instructors School (Pilot 4-Engine)
AAF Pilot School (Specialized 4-Engine)
AAF Weather Station (Type B)
AAF Weather Station (Type A)
Finance Detachment

AAF Weather Station (Type D)
2nd Army Airways Communications Squadron, Detachment
2nd Weather Squadron, Detachment
2nd Weather Squadron, Regional, Detachment (Type D)
69th AAF Base Unit (2nd Weather Region), Section
85th AAF Base Unit, Section H (103rd AACS Squadron), Detachment
703rd AAF Base Unit (Glider Test)
733rd AAF Base Unit (103rd AACS Squadron), Detachment
4143rd AAF Base Unit (Glider Test)
1072nd Guard Squadron, Detachment

Dayton Municipal Airport (III)
Dayton, OH
AAF
Northwest-Vandalia Modification Center #11
702nd AAF Base Unit (Accelerated Service Text)
1072nd Guard Squadron, Detachment

Dayton Army Air Field (III)
Dayton, OH
(Sub Base of Wright Field, OH)
Army Air Forces
Northwest-Vandalia Modification Center #11
Field Office of Materiel Division
502nd AAF Base Unit (HQ, AFATC Field Authorization) - Operating Location #8
Regional Air Priorities Control
ATC Liaison Office
AFATC Liaison Office
Priority Control Office
702nd AAF Base Unit (Accelerated Service Test)
1072nd Guard Squadron, Detachment
4142nd AAF Base Unit (Accelerated Service Test)

Lockbourne Army Air Base (III)
Columbus, OH
Army Service Forces
3568th Service Command Unit: Post Engineers (ASF)
3568th Service Group, Lockbourne Army Air Base (ASF)
Army Air Forces
AAF Combat Crew School
AAF Glider School
Base Weather Station
Glider Pilot Replacement Pool
AAF Instructors School (Pilot 4-Engine)
AAF Pilot School (Specialized 4-Engine)
AAF Weather Station (Type B)
AAF Weather Station (Type A)
Finance Detachment

- Medical Detachment
- Veterinary Detachment
- WAC Detachment
- Third Echelon Repair Shop
- 2nd Army Airways Communications Squadron, Detachment
- 2nd Weather Squadron, Detachment
- 2nd Weather Squadron, Regional, Detachment
- 2nd Weather Squadron, Regional, Detachment (Type A)
- 3rd Airdrome Squadron
- 4th Airdrome Squadron
- 19th Air Depot Group (less Medical Det & Supply Squadron)
- 29th Signal Company, Service Group
- 44th Pilot Transition Training Group (Four-Engine), HQ & HQ Squadron
- 56th Sub-Depot
- 69th AAF Base Unit (2nd Weather Region), Section
- 85th AAF Base Unit, Section H (103rd AACS Squadron), Detachment
- 97th Service Group, HQ & HQ Squadron
- 315th Service Squadron
- 384th Service Squadron
- 101st Quartermaster Company, Service Group (Aviation)
- 374th Base Headquarters & Air Base Squadron
- 394th Aviation Squadron (Colored)
- 415th AAF Band
- 553rd AAF Base Unit (3rd Ferrying Location) - Operation Location #5
- Special Control Office
- 715th AAF Band
- 733rd AAF Base Unit (103rd AACS Squadron), Detachment
- 793rd WAC Post Headquarters Company, AAF
- 856th Signal Service Company, Aviation, Detachment
- 905th Quartermaster Service Company, Aviation, Detachment
- 907th Quartermaster Service Company, Aviation, Detachment
- 907th Specialized Pilot Training Squadron (Four-Engine)
- 908th Specialized Pilot Training Squadron (Four-Engine)
- 909th Specialized Pilot Training Squadron (Four-Engine)
- 912th Specialized Pilot Training Squadron (Four-Engine)
- 1000th Quartermaster Platoon Transport Air Base (Colored)
- 1041st Guard Squadron
- 1077th Ordnance Company, Aviation (Airborne)
- 1172nd Pilot Transition Training Squadron (Four-Engine)
- 1173rd Pilot Transition Training Squadron (Four-Engine)
- 1174th Pilot Transition Training Squadron (Four-Engine)
- 1781st Ordnance Medium Maintenance Company, Aviation (Q)
- 1781st Ordnance Supply & Maintenance Company, Aviation
- 1782nd Ordnance Medium Maintenance Company, Aviation (Q)
- 1782nd Ordnance Supply & Maintenance Company, Aviation
- 1806th Ordnance Medium Maintenance Company, Aviation (Q)
- 1807th Ordnance Medium Maintenance Company, Aviation (Q)
- 2026th Quartermaster Truck Company, Aviation
- 2027th Quartermaster Truck Company, Aviation

2054th Quartermaster Truck Company, Aviation
2055th Quartermaster Truck Company, Aviation
2062nd Ordnance Company, Aviation (Service), Detachment
2114th AAF Base Unit (Pilot School, Specialized 4-Engine & Instructor School)
2114th AAF Base Unit (Pilot School, Specialized 4-Engine & Instructor School), WAC Sqdn

Lunken Airport (III)
Cincinnati, OH

Army Air Forces
Air Support Base
Air Freight Terminal
AAF Flight Service Center
2nd Army Airways Communications Squadron, Detachment
39th Air Freight Wing, Detachment
69th AAF Base Unit (2nd Weather Region), Section
85th AAF Base Unit, Section H (103rd AACS Squadron), Detachment
502nd AAF Base Unit (HQ, AFATC Field Authorization) - Operating Location #31
AFATC Liaison Office
502nd AAF Base Unit (HQ, ATC Field Authorization) - Operating Location
551st AAF Base Unit (Ferrying Division, AFATC, Reserve)
586th AAF Base Unit (Ferrying Squadron, AFATC)
586th AAF Base Unit (Ferrying Squadron, AFATC), WAC Squadron

Ohio State Fairgrounds
Columbus, OH

Army Service Forces
3577th Service Command Unit: 843rd AAF Specialized Depot (ASF)
Medical Detachment (AAF)

Army Air Forces
AAF Storage Depot
3rd AAF Storage Depot
59th Air Depot Group
701st AAF Specialized Depot (AAF)
843rd AAF Specialized Depot (PZ 3)
3577th Service Command Unit Post Engineers, AAF Storage Depot (ASF)
4843rd AAF Base Unit (Specialized Depot)
4843rd AAF Base Unit (Specialized Depot), WAC Squadron

Patterson Field (III)
Osborn, OH

Army Service Forces
Field Printing Plant

Army Air Forces
AAF Regional Hospital
AAF Weather Station (Type A)
AAF Weather Station (Type A-B-S)
Air Freight Terminal, AFATC
AAF Crystal Bank, Headquarters

- AAF Nurses Training Detachment #6
- Air Service Command, HQ & HQ Squadron
- Air Service Command
- Fairfield Air Depot Control Area
- Fairfield Air Depot
- Fairfield Air Service Command, HQ & HQ Squadron
- Fairfield Air Service Command
- Field Printing Plant
- Flying Safety Liaison Branch
- Headquarters, Fairfield Air Technical Service Command
- Finance Detachment #1
- Finance Detachment #23
- Medical Detachment
- Medical Detachment (Colored)
- Veterinary Detachment
- 1st Plant Maintenance Squadron
- 1st Radio Squadron
- 1st Mobile Radio Repair Detachment
- 1st Weather Reconnaissance Squadron
- 1st Radio Squadron
- 2nd Communications Squadron
- 2nd Communications Region, Headquarters
- 2nd Airways Communication Squadron (less Detachments)
- 2nd Airways Communication Region, Headquarters
- 2nd Army Airways Communnications Squadron, Detachment
- 2nd Weather Squadron
- 2nd Weather Region, Headquarters
- 2nd Weather Region
- 2nd Weather Squadron (less Detachments)
- 2nd Weather Service Regional Control Headquarters
- 2nd Weather Squadron, Regional (less Detachments)
- 2nd Weather Squadron, Regional (less Detachments) (Type A-B-S)
- 2nd Weather Squadron, Regional, Detachment
- 4th Air Depot Group, HQ & HQ Squadron
- 4th Repair Squadron
- 4th Supply Squadron
- 10th Transport Group, HQ & HQ Squadron
- 1st Transport Squadron
- 5th Transport Squadron
- 13th Transport Squadron
- 15th Statistical Control Unit (Special), HQ & HQ Det & 15th Reporting Detachment
- 15th Statistical Control Unit (Special), HQ & HQ Det & 9th Reporting Det (Type A)
- 18th Air Depot Group
- 18th Air Depot Group, HQ & HQ Squadron
- 18th Depot Supply Squadron
- 88th Depot Supply Squadron
- 18th Medical Supply Platoon, Aviation
- 19th Air Depot Group - Medical Det & Supply Squadron
- 37th Medical Supply Platoon, Aviation

- 39th Air Freight Wing
- 39th Air Freight Wing (less Detachments)
- 39th Medical Supply Platoon, Aviation
- 45th Medical Supply Platoon, Aviation
- 55th Air Depot Group
- 55th Air Depot Group, HQ & HQ Squadron
- 85th Depot Supply Squadron
- 56th Air Depot Group
- 57th Air Depot Group
- 57th Air Depot Group, HQ & HQ Squadron
- 57th Depot Repair Squadron
- 57th Depot Supply Squadron
- 63rd Transport Group, HQ & HQ Squadron
- 9th Transport Squadron
- 69th AAF Base Unit (2nd Weather Region), (less Sections)
- 69th AAF Base Unit (2nd Weather Region), (less Sections), WAC Squadron
- 70th AAF Base Unit (2nd Weather Region) (less Sections)
- 70th Medical Supply Platoon, Aviation
- 84th Machine Records Unit
- 85th Depot Repair Squadron
- 85th AAF Base Unit, Section H (103rd AACS Squadron), Detachment
- 88th Depot Repair Squadron
- 91st Air Depot Group, HQ & HQ Squadron
- 86th Depot Replacement Squadron
- 94th Depot Supply Squadron
- 97th Depot Supply Squadron
- 98th Aviation Squadron (Separate)
- 98th Aviation Squadron (Colored)
- 308th Service Group, HQ & HQ Squadron
- 308th Service Group, HQ & HQ Squadron
- 315th Depot Repair Squadron
- 345th Aviation Squadron (Colored)
- 351st Aviation Squadron (Colored)
- 361st AAF Band
- 371st Aviation Squadron (Colored)
- 372nd Aviation Squadron (Colored)
- 407th Service Squadron
- 407th Service Squadron (Chinese)
- 418th Quartermaster Platoon (Air Depot Group)
- 419th Quartermaster Platoon (Air Depot Group)
- 437th Quartermaster Platoon, Air Depot Group
- 448th Quartermaster Platoon (Air Depot Group)
- 463rd Aviation Squadron (Colored)
- 478th Base Headquarters & Air Base Squadron
- 661st AAF Band
- 720th Quartermaster Truck Platoon, Aviation (Separate)
- 733rd AAF Base Unit (103rd AACS Squadron), Detachment
- 808th Quartermaster Truck Platoon, Aviation (Separate)
- 836th Guard Squadron

- 859th Signal Service Company, Aviation (less Detachments)
- 895th Signal Company, Depot, Aviation
- 898th Signal Depot Company, Aviation
- 899th Signal Depot Company, Aviation
- 902nd Signal Depot Company, Aviation
- 905th Quartermaster Service Company, Aviation, Detachment
- 905th Quartermaster Service Company, Aviation (less Detachments)
- 912th Engineer Headquarters Company, Air Force
- 918th Signal Company, Depot, Aviation
- 923rd Guard Squadron
- 1408th Quartermaster Company, C1, III, Aviation
- 1465th Ordnance Medium Auto Maintenance Platoon, Aviation
- 1705th Ordnance Medium Maintenance Company, Aviation (Q)
- 1706th Ordnance Medium Maintenance Company, Aviation (Q)
- 1895th Ordnance Supply & Maintenance Co., Aviation (Q)
- 1926th Ordnance Ammunition Company
- 1958th Ordnance Depot Company (Aviation)
- 1972nd Ordnance Depot Company (Aviation)
- 2054th Ordnance Service Company, Aviation (less Detachments)
- 2063rd Quartermaster Truck Company, Aviation
- 2068th Quartermaster Medium Maintenance Company, Aviation
- 2199th Quartermaster Truck Company, Aviation
- 2202nd Quartermaster Truck Company, Aviation
- 2208th Quartermaster Truck Company, Aviation
- 2265th Quartermaster Truck Company, Aviation
- 2467th Quartermaster Truck Company, Aviation
- 2470th Quartermaster Truck Company, Aviation
- 4000th AAF Base Unit (Headquarters Group)
- 4000th AAF Base Unit (Headquarters Group), WAC Section
- 4001st AAF Base Unit (Headquarters, ASC)
- 4015th AAF Base Unit (Area Command)
- 4015th AAF Base Unit (Area Command), WAC Section
- 4100th AAF Base Unit (Air Base)
- 4100th AAF Base Unit (Air Base), WAC Squadron
- 4100th AAF Base Unit (Base Administration)
- 4100th AAF Base Unit (Base Administration), WAC Section
- 4139th AAF Base Unit (Base Administration
- 4900th AAF Base Unit (Aviation Squadron) (Colored)
- 4901st AAF Base Unit (Aviation Squadron) (Colored)
- 4265th AAF Base Unit (Separation Base Provisional)
- 4517th AAF Base Unit (Medical Training)
- 4917th AAF Base Unit (Aviation Squadron) (Colored)

Port Columbus Municipal Airport
Columbus, OH
(Sub Base of Godman Field, KY)
Army Air Force
Air Support Base - S

Vandalia Municipal Airport (III)
Vandalia, OH

Army Air Force
Northwest-Vandalia AAF Modification Center #11
1072nd Guard Squadron, Detachment

Wright Field (III)
Dayton, OH

Army Service Forces
Signal Corps Aircraft Signal Agency (ASF) (CSigO)
Aircraft Radio Laboratory
Aircraft Radio Maintenance Division
Training Film Production Laboratory - Wright Field Branch (ASF)
Aircraft Signal Patent Agency (IV Act)
Dayton Signal Corps Publication Agency (ASF)
3581st Service Command Unit: Station Complement, Signal Corps Aircraft Signal Agency
9390th TSU Ord: Ordnance Section, Wright Field (R & D Service)
9774th TSU CWS: Chemical Warfare Section
9817th TSU CE, Detachment #4: Engineer Board, Aerial Photography Branch

Army Air Forces
Aero-Medical Laboratory
Aircraft Radio Laboratory
Aircraft Radio Maintenance Division
AAF Air Technical Service Command, HQ
AAF Engineer School
AAF Materiel Center
AAF Materiel Command, HQ & HQ Squadron
AAF Materiel Command, Headquarters
AAF Materiel Command, Government Furnished Equipment Depot
AAF Weather Station (Type B)
AAF Weather Station (Type A)
Dayton Signal Corps Publication Agency
Engineer Detachment
Headquarters, Air Technical Service Command
Headquarters, AAF Crystal Bank
Field Printing Plant
Government Furnished Equipment Depot
Medical Detachment
Veterinary Detachment
Signal Aircraft Maintenance Section (SOS)
Signal Aircraft Maintenance Section (ASF)
Signal Corps Aircraft Signal Service (ASF)
Signal Corps Aircraft Signal Agency
Aircraft Radio Maintenance Division
Aircraft Distribution Office
Aircraft Scheduling Unit
Aircraft Radio Laboratry
Training Film Production Laboratory - Wright Field Branch (SOS)

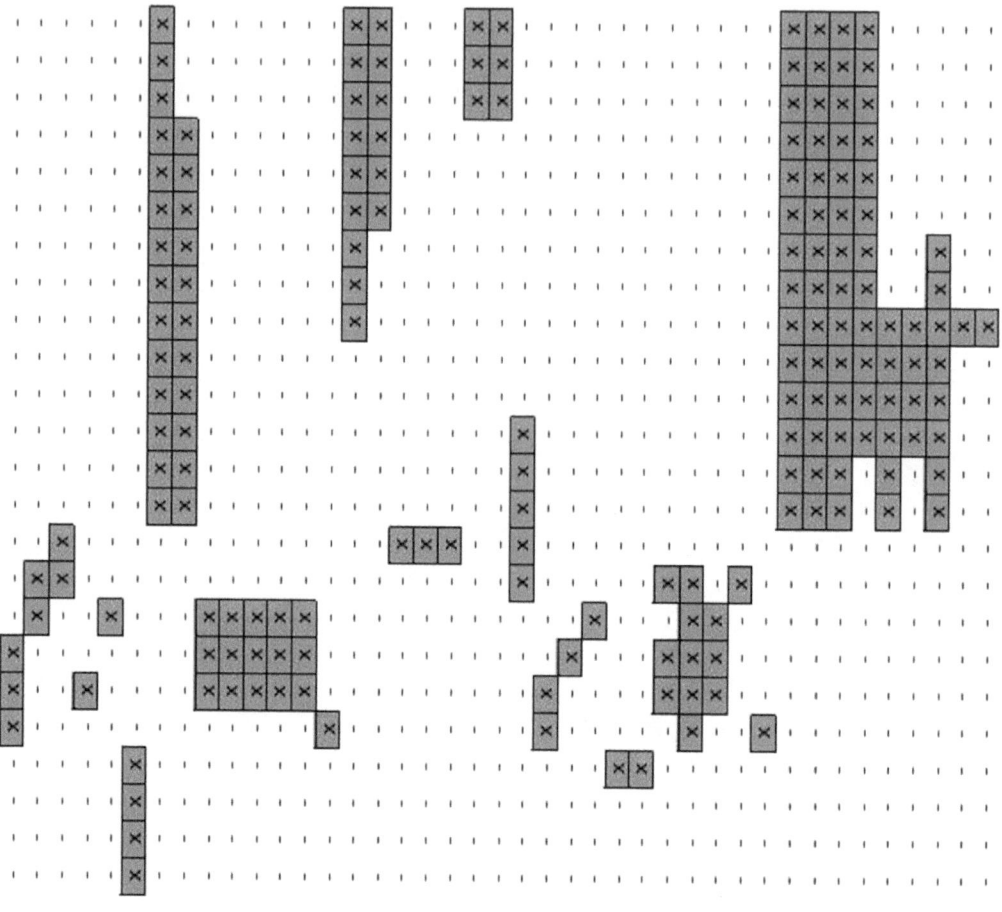

U.S. District Engineer Office (SOS)
2nd Army Airways Communications Squadron, Detachment
2nd Weather Squadron, Detachment
2nd Weather Squadron, Regional, Detachment
2nd Weather Squadron, Regional, Detachment (Type A)
50th Transport Wing, HQ & HQ Squadron
69th AAF Base Unit (2nd Weather Region), Section
85th AAF Base Unit, Section H (103rd AACS Squadron), Detachment
345th Aviation Squadron (Colored)
435th Aviation Squadron (Colored)
436th Aviation Squadron (Colored)
437th Aviation Squadron (Colored)
452nd AAF Band
477th Base Headquarters & Air Base Squadron
553rd AAF Base Unit (3rd Ferrying Group), Operation Location #14
    Area Priorities Office
700th AAF Base Unit (HQ, Materiel Command-Enlisted Personnel)
701st AAF Base Unit (Wright Field Station Complement)
704th AAF Base Unit (HQ, Materiel Command-Officers)
705th Counter Intelligence Corps Detachment
733rd AAF Base Unit (103rd AACS Squadron), Detachment
752nd AAF Band
846th Signal Service Photo Battalion, Company C (SOS)
846th Signal Service Photo Battalion, Company C (ASF)
852nd Signal Service Company, Aviation, Detachmetn #1
854th Ordnance Company (Aviation) (Service)
859th Signal Service Company, Aviation
859th Signal Service Company, Aviation, Detachment #1
1072nd Guard Squadron
1454th Quartermaster Service Company, Aviation
1454th Guard Squadron
2054th Ordnance Service Company, Aviation, Detachment
4000th AAF Base Unit (Command)
4000th AAF Base Unit (Command), WAC Squadron
4020th AAF Base Unit (HQ, ATSC)
4020th AAF Base Unit (HQ, ATSC), WAC Squadron
4141st AAF Base Unit (Air Base)
4141st AAF Base Unit (Air Base), WAC Squadron
4900th AAF Base Unit (Aviation Squadron) (Colored)
9774th TSU CWS: Chemical Warfare Section
9390th TSU Ord: Ordnance Section, Wright Field (R & D Service)

Zanesville Municipal Airport (III)
Zanesville, OH
  AAF

Army Service Forces

Fifth Service Command
Columbus Ohio

Indiana
Army Service Force Stations

Billings General Hospital (I Sub), Fort Benjamin Harrison, Indiana
- Army Service Forces
- Medical Detachment (SOS)
- Medical Technician School (SOS)
- Quartermaster Detachment (SOS)
- Medical Department Officer Replacement Pool
- Medical Enlisted Technicians School
- 22nd WAC Hospital Company (Z)
- 1530th Service Command Unit: Billings General Hospital (incl WAC personnel)
- 1530th Service Command Unit: Medical Department Enlisted Technicians School
- 1530th Service Command Unit: Billings General Hospital
- 1530th Service Command Unit: Separation Point
- 3587th Service Command Unit: Medical Department, Enlisted Technicians School
- 3591st Service Command Unit: Billings General Hospital
- 3591st Service Command Unit: Billings General Hospital, WAC Detachment
- 9950th TSU SGO, Detachment #4: Medical Department Officer Replacement Pool

Casad Ordnance Depot (IV), New Haven, Indiana
- Army Service Forces
- Ordnance Officer Replacement Pool (ASF)
- 3572nd Service Command Unit: Casad Ordnance Depot
- 9315th Technical Service Unit Ordnance: Casad Ordnance Depot

Evansville Ordnance Plant (IV), Evansville, Indiana
- ASF Facility

Fall Creek Ordnance Plant (IV), Indianapolis, Indiana
- ASF Facility

Gary Armor Plate Plant (IV)
- ASF Facility

Hoosier Ordnance Plant (IV), Charlestown, Indiana
- ASF Facility
- 9328th TSU Ordnance, Detachment #21: Hoosier Ordnance Plant

Indiana Ordnance Works (IV), Charlestown, Indiana
- ASF Facility

9328th TSU Ord, Detachment #23: Indiana Ordnance Works

Indianapolis Chemical Warfare Depot (IV), Indianapolis, IN
ASF Facility
3577th Service Command Unit: Indianapolis Chemical Warfare Depot
9731st TSU CWS: Indianapolis Chemical Warfare Depot

Jefferson Proving Ground (IV), Madison, Indiana
Army Service Forces
1559th Service Command Unit: Station Complement
9338th TSU Ord: Jefferson Proving Ground
Army Air Forces
AAF Weather Station (Type D)
Madison Army Air Field
2nd Weather Squadron, Regional - Detachment
2nd Weather Squadron, Detachment
3rd AAF Proving Ground Detachment
69th AAF Base Unit (2nd Weather Region), Section
615th AAF Base Unit (Proving Ground Detachment)
1559th Service Command Unit: Station Complement (SOS)

Jeffersonville Quartermaster Depot (IV), Jeffersonville, Indiana
Army Service Forces
Army Reservation Bureau (ASF)
Quartermaster Procurement District (SOS)
Army Reservation Bureau (ASF)
Field Printing Plant (IV Sub)
Textile Laboratory (ASF)
Quartermaster Officer Replacement Pool
Quartermaster Procurement District (IV Sub)
1502nd Service Command Unit: Finance Office, U. S. Army (SOS)
1561st Service Command Unit: Station Complement
9109th TSU QMC, Detachment #19: Quartermaster Officer Replacement Pool
9151st TSU QMC: Jeffersonville Quartermaster Depot (incl WAC personnel)
Army Air Forces
69th Army Air Force Base Unit (2nd Weather Region) - Section

Kingsbury Ordnance Plant (IV), LaPorte, Indiana
ASF Facility
9328th TSU Ord, Detachment #30: Kingsbury Ordnance Plant

Camp Thomas A. Scott (I), Fort Wayne, Indiana
Army Service Forces
ASF Facility
717th Railway Operations Battalion, Transportation Corps (ASF)

- 745th Railway Operations Battalion, Transportation Corps (ASF)
- 750th Railway Operations Battalion, Transportation Corps (ASF)
- 1564th Service Command Unit: Station Complement (SOS)

Terre Haute Ordnance Depot (IV), Terre Haute, Indiana
Army Service Forces
- ASF Facility
- 3573rd Service Command Unit: Terre Haute Ordnance Depot
- 9383rd TSU Ordnance: Terre Haute Ordnance Depot

Vigo Plant, Chemmical Warfare Service (IV), Terre Haute, IN
Army Service Forces
- ASF Facility
- Chemical Warfare Service Detachment
- 1517th Service Command Unit: Vigo Plant, Chemical Warfare Service
- 1517th Service Command Unit: Vigo Plant, Headquarters Section
- 9768th TSU CWS: Vigo Plant, Chemical Warfare Service
- 9768th TSU CWS: Vigo Plant, Chemical Warfare Service Detachment
- 9768th TSU CWS: Vigo Plant, Military Police Detachment
- 9768th TSU CWS: Vigo Plant, Station Hospital

Wabash River Ordnance Works (IV), Dana, Indiana
- ASF Facility
- 9328th TSU Ordnance, Detachment #48: Wabash River Ordnance Works

Wakeman General Hospital (I Act), Camp Atterbury, Indiana
Army Service Forces
- ASF Convalescent Hospital
- Wakeman Convalescent Hospital (ZI)
- Wakeman General Hospital
- Medical Department School
- Medical Department Enlisted Technicians School (WAC)
- 21st WAC Hospital Company (Colored) (ZI)
- 54th WAC Hospital Company (ZI)
- 1560th Service Command Unit: Medical Dept. Enlisted Technicians School (WAC)
- 1560th Service Command Unit: Wakeman Gen & Conval Hosp (incl WAC personnel)
- 1560th Service Command Unit: Wakeman Hospital Center
- 1560th Service Command Unit: Separation Point
- 3545th Service Command Unit: Medical Dept Enlisted Technicians School (WAC)
- 3547th Service Command Unit: Wakeman General & Convalescent Hospital
- 3547th Service Command Unit: Wakeman General & Conval Hospital, WAC Detachment
- 3547th Service Command Unit: Wakeman General & Conval Hospital, WAC Det (Colored)

War Department Personnel Center, Camp Atterbury, Indiana
Army Service Forces

Reception Center
Reception Station #6
1534th Service Command Unit: Reception Center (ASF)
1560th Service Command Unit: War Department Personnel Center
Reception Center
Reception Station #6
Separation Center
Separation Center #31
Casual Section
Special Training Unit
1558th Service Command Unit: Reception Station #6
1585th Service Command Unit: Separation Center

Kentucky
Army Service Force Stations

Blue Grass Ordnance Depot (IV), Richmond, Kentucky
ASF Facility
3570th Service Command Unit: Blue Grass Ordnance Depot (ASF)
9310th Technical Service Unit: Blue Grass Ordnance Depot

Darnall General Hospital (I), Danville, Kentucky
ASF Facility
Army Service Forces
Medical School - Medical Training Detachment (SOS)
Medical School Detachment (ASF)
355th ASF Band
1591st Service Command Unit: Station Complement (SOS)
3527th Service Command Unit: Casual Company (ASF)
3592nd Service Command Unit: Darnall General Hospital
3592nd Service Command Unit: Separation Point

Kentucky Ordnance Works (IV), Paducah, Kentucky
ASF Facility
9328th TSU Ord, Detachment #28: Kentucky Ordnance Works

Lexington Signal Depot (IV), Avon, Kentucky
Army Service Forces
Signal Corps Officer Replacement Pool (SOS)
Signal Corps Supply Officer School (ASF)
Lexington Signal Depot Warehouse (IV Sub) (Old Burley)
Lexington Signal Depot Warehouse (IV Sub) (New Burley)
Post Photographic Laboratory (ASF)
Signal Corps Photographic Laboratory (ASF)

- 188th Signal Repair Company (ASF)
- 189th Signal Repair Company (ASF)
- 1549th Service Command Unit: Station Complement
- 1549th Service Command Unit: Lexington Signal Depot
- 9504th TSU Signal Corps: Lexington Signal Depot (incl WAC personnel)

Army Ground Forces
- 55th Signal Repair Company
- 58th Signal Repair Company

Louisville Medical Depot (IV), Louisville, Kentucky
Army Service Forces
- 3574th Service Command Unit: Louisville Medical Depot
- 9343rd TSU Ord: Ordnance Section
- 9909th TSU SGO: Louisville Medical Depot Fiscal Activities

Louisville Ordnance Depot (IV), Louisville, Kentucky
- ASF Facility

Nichols General Hospital (I), Louisville, Kentucky
Army Service Forces
- Service School for Medical Officers (ASF)
- Medical Department School
- 18th WAC Hospital Company (ZI)
- 25th General Hospital (SOS)
- 459th ASF Band
- 1572nd Service Command Unit: US Army Recruiting Station
- 1572nd Service Command Unit: US Army WAC Recruiting Station
- 3530th Service Command Unit: Casual Company (ASF)
- 3595th Service Command Unit: Nichols General Hospital
- 3595th Service Command Unit: Nichols General Hospital, WAC Detachment
- 3595th Service Command Unit: Nichols General Hospital (incl WAC personnel)
- 3595th Service Command Unit: Separation Point

Ohio River Ordnance Works (M), Henderson, Kentucky
- ASF Facility
- 9328th TSU Ordnance, Detachment #39: Ohio Rover Ordnance Works

War Department Personnel Center (I Act), Fort Knox, Kentucky
Army Service Forces
- 1550th Service Command Unit: War Department Personnel Center
- 1550th Service Command Unit: Separation Center #48
- Separation Center Section
- 1550th Service Command Unit: Reception Station #24

Ohio
Army Service Force Stations

Buckeye Ordnance Works (IV), Ironton, Ohio
- ASF Facility
- 9326th Technical Service Unit Ordnance, Detachment #4: Buckeye Ordnance Works
- 9328th Technical Service Unit Ordnance, Detachment #6: Buckeye Ordnance Works

Bucyrus Transportation Corps Railroad Repair Shop (IV Sub), Bucyrus, Ohio
- ASF Facility

Cambridge Engineer Sub Depot (IV Sub), Cambridge, OH
- ASF Facility

Columbus Adjutant General Depot (I), Columbus, Ohio
- Army Service Forces
- Field Printing Plant
- 1541st Service Command Unit: Columbus Adjutant General Depot
- 1541st Service Command Unit: Station Complement (ASF)

Columbus Quaratermaster Depot, Columbus, Ohio
- Engineer Supply School (SOS)
- Habus CWS Plant (SOS)
- Medical School - Medical Training Detachment (SOS)
- 463rd Engineer DepotCompany (SOS)
- 1529th Service Command Unit Station Complement (SOS)

Columbus Army Service Forces Depot (IV), Columbus, OH
- Army Service Forces
- Habus Chemical Warfare Service Plant (IV Act)
- Chemical Warfare Renovating Section
- Engineer Supply School (ASF)
- Quartermaster Packaging and Packing Course (IV Act)
- Engineer Section (IV Sub)
- Medical Section (IV Sub)
- Medical School Detachment (ASF)
- Post Photographic Laboratory (ASF)
- Quartermaster Officer Replacement Pool
- Instrument Repair School
- US District Engineer Office (ASF)
- 681st Engineer Depot Company - Parts Supply Battalion
- 732nd Engineer Depot Company, Detachment
- 752nd Engineer Parts Supply Company (ASF)
- 753rd Engineer Parts Supply Company (ASF)

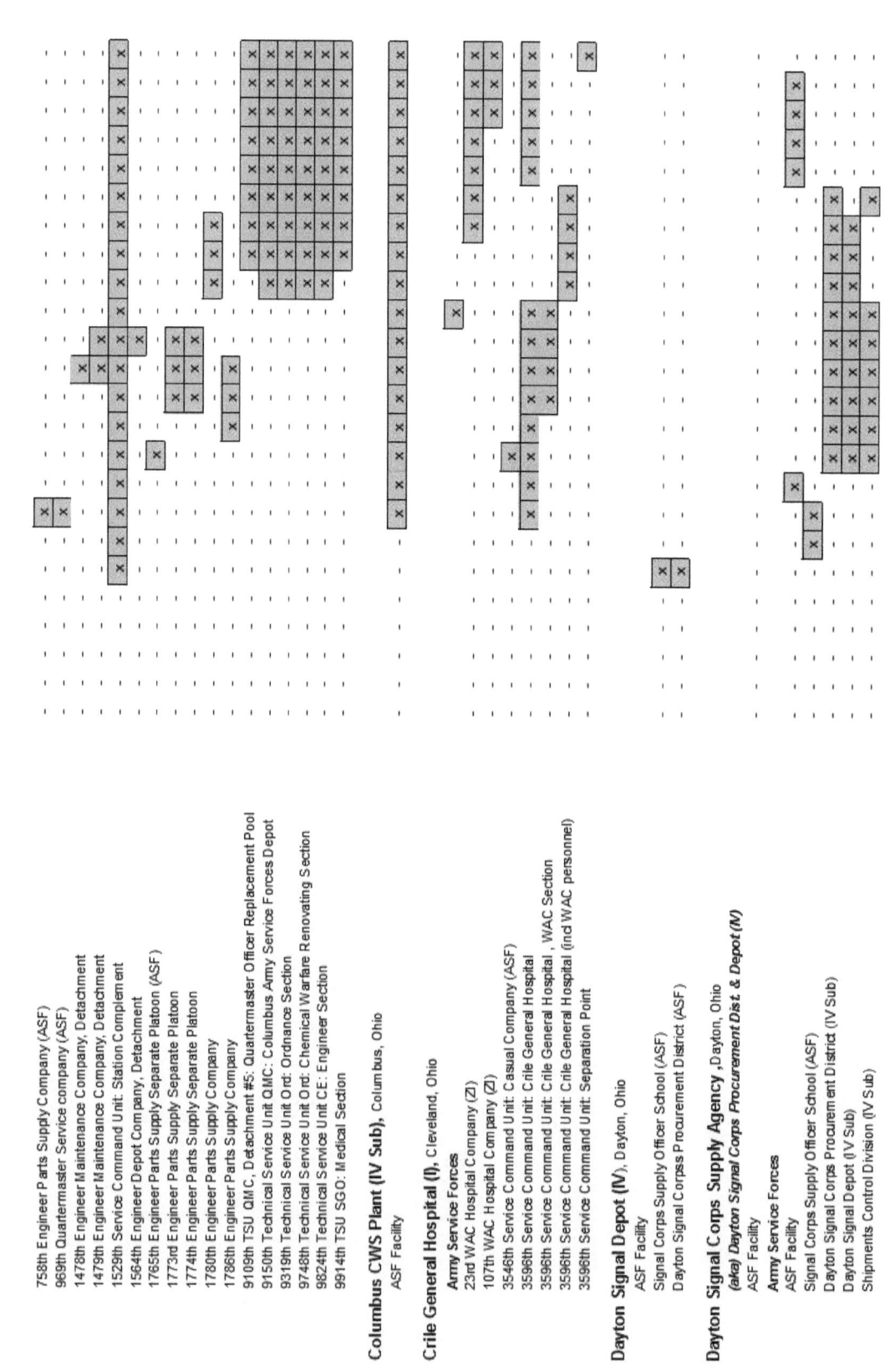

53rd Italian Quartermaster Service Company
1566th Service Command Unit: Dayton Signal Corps Supply Agency
9500th TSU Signal Corps: Dayton Shipments Control Division
9502nd TSU Signal Corps: Dayton Signal Depot (incl WAC personnel)
9530th TSU Signal Corps: Dayton Procurement District (Procurement Activity)
9531st TSU Signal Corps: Dayton Procurement District
9550th TSU Signal Corps: Dayton Shipments Control Division
9551st TSU Signal Corps: Dayton Signal Corps Supply Agency HQ (incl WAC prsnl)

Army Ground Forces
556th Signal Depot Company (AGF)

Army Air Force
Field Office Air Service Division, Supply Maintenance Branch

Erie Ordnance Depot, LaCarne, Ohio

Army Service Forces
84th Ordnance Company (Depot)

Erie Proving Ground (Ordnance) (IV), LaCarne, Ohio

Army Service Forces
Erie Ordnance Depot (SOS)
Erie Regional Office, Cleveland Ordnance District (IV Sub)
Ordnance Officer Replacement Pool (ASF)
50th Italian Quartermaster Service Company
92nd Italian Quartermaster Service Company
93rd Italian Quartermaster Service Company
94th Italian Quartermaster Service Company
101st Italian Quartermaster Service Company
121st Italian Quartermaster Service Company
304th Ordnance Base Regiment, HQ & HQ Detachment, 4th Battalion & Company N
323rd Italian Quartermaster Battalion, HQ & HQ Detachment
323rd Italian Quartermaster Battalion, HQ & HQ Det watchd Medical
1571st Service Command Unit: Station Complement
  Reconditioning Program Section
1571st Service Command Unit: Erie Proving Ground
9327th TSU Ord: Erie Proving Ground, Detachment #1, Erie Ordnance Depot
9327th TSU Ord: Detachment #1, Erie Ordnance Depot

Fern Bank Engineer Depot, Fern Bank, Ohio

ASF (SOS) Facility

Firelands CWS Plant (IV Sub), Marion, Ohio

(aka) Scioto Ordnance Plant

ASF Facility

Status

(Sub Installation of the Pittsburgh Chemical Warfare Procurement District)

Fletcher General Hospital (I), Cambridge, Ohio
- **Army Service Forces**
  - Medical Detachment
  - Medical Department School
  - Quartermaster Detachment
  - 19th WAC Hospital Company (Z)
  - 108th WAC Hospital Company (Z)
  - 1516th Service Command Unit: US Army WAC Recruiting Station
  - 3528th Service Command Unit: Casual Company (ASF)
  - 3593rd Service Command Unit: Fletcher General Hospital
  - 3593rd Service Command Unit: Fletcher General Hospital, WAC Detachment
  - 3593rd Service Command Unit: Fletcher General Hospital (incl WAC personnel)
  - 3593rd Service Command Unit: Separation Point

Fostoria CWS Plant #1 (IV Sub), Fostoria, Ohio
- **Status**
  - (Sub Installation of the Pittsburgh Chemical Warfare Procurement District)
- ASF Facility

Habus Chemical Warfare Service Plant (IV Act), Columbus, OH
(aka) Columbus Army Service Forces Depot
- **Status**
  - ASF Facility

Kings Mills Ordnance Plant (IV), King Mills, Ohio
- ASF Facility

Lima Tank Depot (IV), Lima, Ohio
- ASF Facility

Lordstown Ordnance Depot (IV), Lordstown, Ohio
- ASF Facility
- 3571st Service Command Unit: Lordstown Ordnance Depot (ASF)
- 9341st TSU Ord: Lordstown Ordnance Depot

Marion Engineer Depot (IV), Marion, Ohio
- **Army Service Forces**
  - Engineer Officer Replacement Pool (ASF)
  - 710th Engineer Base Depot Company (ASF)
  - 724th Engineer Base Depot Company (ASF)
  - 3576th Service Command Unit: Marion Engineer Depot
  - 9905th TSU CE, Detachment #6: Marion Engineer Depot

Marion Quartermaster Depot (IV), Marion, Ohio

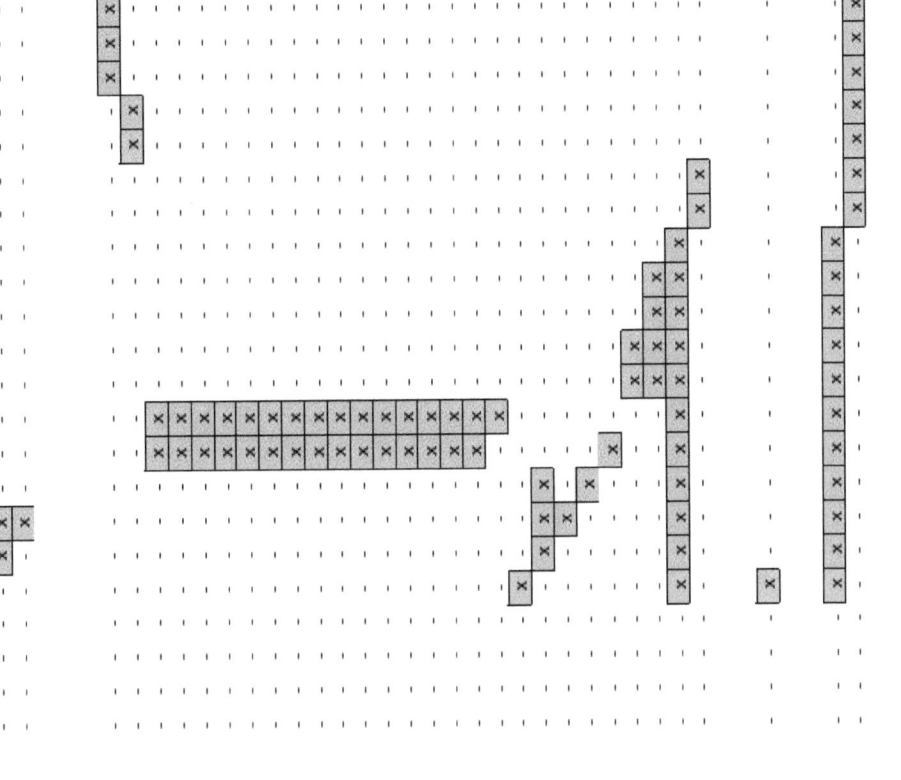

Ravenna Ordnance Center (IV), Ravenna, Ohio
(aka) Portage Ordnance Depot (IV Sub)
(aka) Ravenna Ordnance Plant (IV Sub)
Army Service Forces
    Portage Ordnance Depot (IV Sub)
        1569th Service Command Unit: Station Complement (ASF)
    Ravenna Ordnance Plant (IV Sub)
    9360th TSU Ordnance: Ravenna Ordnance Center (Field Service Depot)
        Detachment #1: Ravenna Ordnance Center (Industrial Service Activities)

Ravenna Ordnance Plant (IV Sub), Ravenna, Ohio
(aka) Ravenna Ordnance Center (IV)
    ASF Facility

Rossford Ordnance Depot (IV), Toledo, Ohio
Army Service Forces
    SOS Facility
    Ordnance Parts Clerical School (IV Sub)
        Provisional Company A
        Provisional Company B
        Provisional Company C
    99th Italian Quartermaster Service Company
    100th Italian Quartermaster Service Company
    101st Italian Quartermaster Service Company
    310th Italian Quartermaster Battalion, HQ & HQ Detachment watchd Medical
    1543rd Service Command Unit: Parts Clerk Course, Rossford Ord Depot, (Station Com)
    1543rd Service Command Unit: Parts Clerk Course, Rossford Ord Depot
    1543rd Service Command Unit: Station Complement, Ordnance Parts Clerical School
    9365th TSU Ordnance: Rossford Ordnance Depot

Scioto Ordnance Plant (IV), Marion, Ohio
    SOS Facility
Army Service Forces
    Firelands Chemical Warfare Service Plant (ASF) (IV)
Army Air Forces
    838th AAF Specialized Depot, Detachment (AAF)

Sharonville Engineer Depot (IV), Sharonville, Ohio
Army Service Forces
    443rd Engineer Base Depot Company (ASF)
    725th Engineer Base Depot Company (ASF)
    1387th Engineer Base Depot Company
    3575th Service Command Unit: Sharonville Engineer Depot
    9805th TSU CE, Detachment #7: Sharonville Engineer Depot

Springfield Engineer Sub Depot (IV Sub), Springfield, Ohio
  ASF Facility

Toledo Medical Supply Depot (IV), Toledo, Ohio
  ASF Facility
  3579th Service Command Unit: Toledo Medical Depot (ASF)
  9912th TSU SGO: Toledo Medical Supply

Youngstown Quartermaster Depot (IV), Youngstown, Ohio
  ASF Facility

Zanesville CWS Plant (IV Sub), Zanesville, Ohio
  ASF Facility

West Virginia
Army Service Force Stations

Ashford General Hospital, White Sulphur Springs, West Virginia
  Army Service Forces
  Ashford Enemy Alien Internment Camp (SOS)
  Medical Detachment (ASF)
  Quartermaster Detachament (SOS)
  Field Printing Plant (IV Act)
  Service School for Medical Officers
  52nd WAC Hospital Company (ZI)
  53rd WAC Hospital Company (ZI)
  453rd ASF Band
  507th AAF Band
  1573rd Service Command Unit: US Army Recruiting Station
  1573rd Service Command Unit: US Army WAC Recruiting Station
  3526th Service Command Unit: Casual Company (ASF)
  3590th Service Command Unit: Ashford General Hospital
  3590th Service Command Unit: Ashford General Hospital, POW Camp
  3590th Service Command Unit: Ashford General Hospital, Separation Point

Newton D. Baker General Hospital, Vanclevesville, West Virginia
  Army Service Forces
  ASF Facility
  20th WAC Hospital Company (ZI)
  106th WAC Hospital Company (ZI)
  3529th Service Command Unit: Casual Company (ASF)
  3594th Service Command Unit: Newton D. Baker General Hospital
  3594th Service Command Unit: Separation Point

Kanawha CWS Plant (IV Sub), South Charleston, West Virginia
  ASF Facility
  9326th TSU Ordnance, Detachment #36: Morgantown Ordnance Works
  Surplus

Marietta Manufacturing Plant, Point Pleasant, West Virginia
  ASF Facility

Marshall CWS Plant (IV Sub), New Martinville, West Virginia
  ASF Facility

Morgantown Ordnance Works (IV), Morgantown, West Virginia
  ASF Facility
  9328th TSU Ordnance, Detachment #36: Morgantown Ordnance Works

West Virginia Ordnance Works (IV), Point Pleasant, West Virginia
  ASF Facility
  9328th TSU Ordnance: Ordnance Section, Wright Field (R & D Service)
  9328th TSU Ord, Detachment #50: West Virginia Ordnance Works

———————————————— POW

Fifth Service Command

Prisoner of War Camps

Camp Atterbury Internment Camp
Camp Atterbury, IN

Army Service Forces

Unit	1941-5	1941-7	1941-8	1941-11	1943-1	1943-6	1943-9	1943-12	1944-3	1944-6	1944-9	1944-10	1944-11	1944-12	1945-1	1945-2	1945-3	1945-4	1945-5	1945-6	1945-7	1945-8	1945-9	1945-10
414th Military Police Escort Guard Company (less Det) (ASF)	-	-	-	-	x	-	-	-	-	-	-	-	-	-	-	-	-	-	-	-	-	-	-	-
428th Military Police Escort Guard Company (ASF)	-	-	-	-	x	-	-	-	-	-	-	-	-	-	-	-	-	-	-	-	-	-	-	-
429th Military Police Escort Guard Company (ASF)	-	-	-	-	x	-	-	-	-	-	-	-	-	-	-	-	-	-	-	-	-	-	-	-
1537th Service Command Unit: Station Complement (ASF)	-	-	-	-	x	-	-	-	-	-	-	-	-	-	-	-	-	-	-	-	-	-	-	-

Camp Atterbury POW Camp
Camp Atterbury, IN

Army Service Forces

Unit	1941-5	1941-7	1941-8	1941-11	1943-1	1943-6	1943-9	1943-12	1944-3	1944-6	1944-9	1944-10	1944-11	1944-12	1945-1	1945-2	1945-3	1945-4	1945-5	1945-6	1945-7	1945-8	1945-9	1945-10
414th Military Police Escort Guard Company (less Det) (ASF)	-	-	-	-	-	x	x	-	-	-	-	-	-	-	-	-	-	-	-	-	-	-	-	-
428th Military Police Escort Guard Company (ASF)	-	-	-	-	-	x	x	-	-	-	-	-	-	-	-	-	-	-	-	-	-	-	-	-
429th Military Police Escort Guard Company (ASF)	-	-	-	-	-	x	x	-	-	-	-	-	-	-	-	-	-	-	-	-	-	-	-	-
577th Military Police Escort Guard Company (ASF)	-	-	-	-	-	-	x	x	-	-	-	-	-	-	-	-	-	-	-	-	-	-	-	-
1537th Service Command Unit: Prisoner of War Camp	-	-	-	-	-	x	x	x	x	-	-	-	-	-	-	-	-	-	-	-	-	-	-	-
1560th Service Command Unit: Prisoner of War Camp	-	-	-	-	-	-	-	-	x	x	x	x	x	x	x	x	x	x	x	x	-	-	-	-
Austin Branch POW Camp (I Sub)	-	-	-	-	-	-	-	-	-	-	x	x	-	-	-	-	-	-	-	-	-	-	-	-
1560th Service Command Unit: Prisoner of War Camp	-	-	-	-	-	-	-	-	-	-	-	-	x	x	x	x	x	x	x	x	x	x	x	-
1560th Service Command Unit: Detachment, Prisoner of War Camp	-	-	-	-	-	-	-	-	-	-	-	-	x	x	x	x	x	x	x	x	x	-	-	-
Eaton Branch POW Camp	-	-	-	-	-	-	-	-	-	-	-	-	-	-	-	-	-	-	-	-	-	-	-	-
1560th Service Command Unit: Detachment, Prisoner of War Camp	-	-	-	-	-	-	-	-	-	-	-	-	-	-	-	-	-	-	-	x	x	-	-	-
Morristown Branch POW Camp	-	-	-	-	-	-	-	-	-	-	-	-	-	-	-	-	-	-	-	-	-	-	-	-
1560th Service Command Unit: Detachment, Prisoner of War Camp	-	-	-	-	-	-	-	-	-	-	-	-	-	-	-	-	-	-	-	x	-	-	-	-
Vincennes Branch POW Camp	-	-	-	-	-	-	-	-	-	-	-	-	-	-	-	-	-	-	-	-	-	-	-	-
1560th Service Command Unit: Detachment, Prisoner of War Camp	-	-	-	-	-	-	-	-	-	-	-	-	-	-	-	-	-	-	-	x	-	-	-	-
Windfall Branch POW Camp	-	-	-	-	-	-	-	-	-	-	-	-	-	-	-	-	-	-	-	-	-	-	-	-
1560th Service Command Unit: Detachment, Prisoner of War Camp	-	-	-	-	-	-	-	-	-	-	-	-	-	-	-	-	-	-	-	-	-	x	x	-

Fort Benjamin Harrison POW Camp (I Sub)
Fort Benjamin Harrison, IN

Army Service Forces

Unit	1941-5	1941-7	1941-8	1941-11	1943-1	1943-6	1943-9	1943-12	1944-3	1944-6	1944-9	1944-10	1944-11	1944-12	1945-1	1945-2	1945-3	1945-4	1945-5	1945-6	1945-7	1945-8	1945-9	1945-10
3570th Service Command Unit: Prisoner of War Camp	-	-	-	-	-	-	-	-	x	x	x	x	x	-	-	-	-	-	-	-	-	-	-	-
1530th Service Command Unit: Prisoner of War Camp	-	-	-	-	-	-	-	-	-	-	-	-	-	-	x	-	-	-	-	-	-	-	-	-

* - Same facility reflecting a change in the name (when "bound" together).

Fifth Service Command
Columbus Ohio

Maneuver Areas

Tennessee Maneuver Area
Franklin, Kentucky

Army Ground Forces

Tennessee Maneuver Area
Petroleum, KY

Army Ground Forces

Tennessee Maneuver Area
Scottsville, KY

Army Service Forces

Army Ground Forces

Unit	1941				1942		1943				1944							1945									
	5	7	8	11	1	6	1	6	9	12	1	3	6	9	10	11	12	1	2	3	4	5	6	7	8	9	10
334th Ordnance Depot Company (AGF)	-	-	-	-	-	-	-	-	X	-	-	-	-	-	-	-	-	-	-	-	-	-	-	-	-	-	-
100th Ordnance Ammunition Battalion, HQ & HQ Detachment (AGF) (Colored)	-	-	-	-	-	-	-	X	X	X	-	-	-	-	-	-	-	-	-	-	-	-	-	-	-	-	-
574th Ordnance Ammunition Battalion, HQ & HQ Detachment (AGF) (Colored)	-	-	-	-	-	-	-	X	X	X	-	-	-	-	-	-	-	-	-	-	-	-	-	-	-	-	-
616th Ordnance Ammunition Battalion, HQ & HQ Detachment (AGF)	-	-	-	-	-	-	-	X	X	X	-	-	-	-	-	-	-	-	-	-	-	-	-	-	-	-	-
618th Ordnance Ammunition Battalion, HQ & HQ Detachment (AGF)	-	-	-	-	-	-	-	X	X	X	-	-	-	-	-	-	-	-	-	-	-	-	-	-	-	-	-
111th Army Postal Unit (atchd AGF)	-	-	-	-	-	-	-	-	-	-	-	X	X	-	-	-	-	-	-	-	-	-	-	-	-	-	-
24th Evacuation Hospital, Semi Mobile (AGF)	-	-	-	-	-	-	-	X	X	X	X	X	-	-	-	-	-	-	-	-	-	-	-	-	-	-	-
35th Evacuation Hospital, Semi Mobile (AGF)	-	-	-	-	-	-	-	-	X	X	X	X	-	-	-	-	-	-	-	-	-	-	-	-	-	-	-
126th Ordnance Medium Maintenance Company (AGF)	-	-	-	-	-	-	-	X	X	X	X	-	-	-	-	-	-	-	-	-	-	-	-	-	-	-	-
169th Ordnance Ammunition Battalion, HQ & HQ Detachment (AGF) (Colored)	-	-	-	-	-	-	-	X	X	X	X	-	-	-	-	-	-	-	-	-	-	-	-	-	-	-	-
265th Ordnance Light Maintenance Company (AGF)	-	-	-	-	-	-	-	X	X	X	-	-	-	-	-	-	-	-	-	-	-	-	-	-	-	-	-
341st Medical Regiment (less Band) (AGF)	-	-	-	-	-	-	-	X	X	X	-	-	-	-	-	-	-	-	-	-	-	-	-	-	-	-	-
341st Medical Group, HQ & HQ Detachment (AGF)	-	-	-	-	-	-	-	-	X	X	-	-	-	-	-	-	-	-	-	-	-	-	-	-	-	-	-
183rd Medical Battalion, HQ & HQ Detachment (Separate)	-	-	-	-	-	-	-	-	-	X	X	X	-	-	-	-	-	-	-	-	-	-	-	-	-	-	-
473rd Medical Collecting Company (Separate)	-	-	-	-	-	-	-	-	-	X	X	X	-	-	-	-	-	-	-	-	-	-	-	-	-	-	-
474th Medical Collecting Company (Separate)	-	-	-	-	-	-	-	-	-	X	X	X	-	-	-	-	-	-	-	-	-	-	-	-	-	-	-
475th Medical Collecting Company (Separate)	-	-	-	-	-	-	-	-	-	X	X	X	-	-	-	-	-	-	-	-	-	-	-	-	-	-	-
625th Clearing Company (Medical) (Separate)	-	-	-	-	-	-	-	-	-	X	X	X	-	-	-	-	-	-	-	-	-	-	-	-	-	-	-
666th Clearing Company (Medical) (Separate)	-	-	-	-	-	-	-	-	-	X	X	-	-	-	-	-	-	-	-	-	-	-	-	-	-	-	-
184th Medical Battalion, HQ & HQ Detachment (Separate)	-	-	-	-	-	-	-	-	-	X	X	X	-	-	-	-	-	-	-	-	-	-	-	-	-	-	-
476th Medical Collecting Company (Separate)	-	-	-	-	-	-	-	-	-	X	X	X	-	-	-	-	-	-	-	-	-	-	-	-	-	-	-
477th Medical Collecting Company (Separate)	-	-	-	-	-	-	-	-	-	X	X	X	-	-	-	-	-	-	-	-	-	-	-	-	-	-	-
478th Medical Collecting Company (Separate)	-	-	-	-	-	-	-	-	-	X	X	X	-	-	-	-	-	-	-	-	-	-	-	-	-	-	-
626th Clearing Company (Medical) (Separate)	-	-	-	-	-	-	-	-	-	X	X	X	-	-	-	-	-	-	-	-	-	-	-	-	-	-	-
349th Ordnance Depot Company (AGF)	-	-	-	-	-	-	-	-	-	-	X	-	-	-	-	-	-	-	-	-	-	-	-	-	-	-	-
877th Ordnance Heavy Auto Maintenance company (AGF)	-	-	-	-	-	-	-	-	-	-	X	-	-	-	-	-	-	-	-	-	-	-	-	-	-	-	-
3420th Ordnance Medium Auto Maintenance Company (AGF) (Colored)	-	-	-	-	-	-	-	-	-	-	X	-	-	-	-	-	-	-	-	-	-	-	-	-	-	-	-
3534th Ordnance Medium Auto Maintenance Company (AGF)	-	-	-	-	-	-	-	-	-	-	X	-	-	-	-	-	-	-	-	-	-	-	-	-	-	-	-

	1941			1942		1943			1944				1945									
	5	7	8	11	1	6	1	6	9	12	3	6	9	10	11	12	1	2	3	4	5	6	7	8	9	10

Tennessee Maneuver Area
Lebanon, TN

Tennessee Maneuver Area
Manchester, TN

Tennessee Maneuver Area
Shelbyville, TN

Tennessee Maneuver Area
Wartrace, TN

Tennessee Maneuver Area
Westmoreland, TN

West Virginia Maneuver Area
Elkins, WV — x x —

Camp Breckinridge Internment Camp (I)
Camp Breckinridge, KY
Army Service Forces
  405th Military Police Escort Guard Company (ASF)
  406th Military Police Escort Guard Company (ASF)
  440th Military Police Company (POW Processing) (ASF)
  1538th Service Command Unit: Station Complement (ASF)

Camp Breckinridge POW Camp
Camp Breckinridge, KY
Army Service Forces
  405th Military Police Escort Guard Company (ASF)
  406th Military Police Escort Guard Company (ASF)
  646th Military Police Escort Guard Company (ASF)
  1538th Service Command Unit: Prisoner of War Camp
  1570th Service Command Unit: Prisoner of War Camp
  Owensboro Branch POW Camp
  1570th Service Command Unit: Detachment, Prisoner of War Camp

Camp Campbell Internment Camp (I)
Clarksville, TN
Army Service Forces
  1539th Service Command Unit: Station Complement (ASF)

Camp Campbell POW Camp
Camp Campbell, KY
Army Service Forces
  421st Military Police Escort Guard Company (ASF)
  494th Military Police Escort Guard Company (ASF)
  658th Military Police Escort Guard Company (ASF)
  1539th Service Command Unit: Prisoner of War Camp
  1580th Service Command Unit: Prisoner of War Camp

Fort Knox POW Camp (I Act)
Fort Knox, KY
Army Service Forces
  414th Military Police Escort Guard Company (ASF)
  3571st Service Command Unit: Prisoner of War Camp
  1550th Service Command Unit: Prisoner of War Camp
  Danville Branch POW Camp
    1550th Service Command Unit: Detachment, Prisoner of War Camp
  Darnall General Hospital Branch POW Camp *(replaces Danville, KY)*
    1550th Service Command Unit: Detachment, Prisoner of War Camp
  Eminence Branch POW Camp
  Frankfort Branch POW Camp
    1550th Service Command Unit: Detachment, Prisoner of War Camp

- Indiana Ordnance Works Branch POW Camp
- 1550th Service Command Unit: Detachment, Prisoner of War Camp
- Jeffersonville (IN) Branch POW Camp (I Sub)
- 1550th Service Command Unit: Prisoner of War Camp
- 1550th Service Command Unit: Detachment, Prisoner of War Camp
- Lexington Branch POW Camp
- 1550th Service Command Unit: Detachment, Prisoner of War Camp
- Lexington Signal Depot Branch POW Camp
- 1550th Service Command Unit: Detachment, Prisoner of War Camp
- Louisville Branch POW Camp
- 1550th Service Command Unit: Detachment, Prisoner of War Camp
- Maysville Branch POW Camp
- Paris Branch POW Camp
- 1550th Service Command Unit: Detachment, Prisoner of War Camp
- Shelbyville (KY) Branch POW Camp
- 1550th Service Command Unit: Detachment, Prisoner of War Camp

Camp Perry POW Camp (I Sub)
Camp Perry, OH

Army Service Forces

- 408th Military Police Escort Guard Company (ASF)
- 412th Military Police Escort Guard Company (ASF)
- 729th Military Police Escort Guard Company (less Cos B & D) (ASF)
- 1590th Service Command Unit: Station Complement (ASF)
- 1590th Service Command Unit: Prisoner of War Camp
- 1591st Service Command Unit: Prisoner of War Camp
- Bowling Green Branch POW Camp
- 1590th Service Command Unit: Detachment, Prisoner of War Camp
- Celina Branch POW Camp
- 1590th Service Command Unit: Detachment, Prisoner of War Camp *(replaces Fort Hayes POW Camp)*
- Columbus Branch POW Camp
- 1590th Service Command Unit: Detachment, Prisoner of War Camp
- Crile General Hospital Branch POW Camp
- 1590th Service Command Unit: Prisoner of War Camp
- 1590th Service Command Unit: Detachment, Prisoner of War Camp
- 107th WAC Hospital Company (ZI)
- Defiance Branch POW Camp
- 1590th Service Command Unit: Detachment, Prisoner of War Camp
- Fort Wayne Branch POW Camp *(aka Camp Thomas A. Scott)*
- 1590th Service Command Unit: Detachment, Prisoner of War Camp
- 1590th Service Command Unit: Detachment, Prisoner of War Camp
- Fletcher General Hospital POW Camp
- 1590th Service Command Unit: Prisoner of War Camp
- 1590th Service Command Unit: Detachment, Prisoner of War Camp
- Fort Hayes Branch POW Camp
- 1590th Service Command Unit: Detachment, Prisoner of War Camp

Marion Branch Camp
  1590th Service Command Unit: Prisoner of War Camp
  1590th Service Command Unit:Detachment, Prisoner of War Camp
Wilmington Branch POW Camp
  1590th Service Command Unit:Detachment, Prisoner of War Camp

Ashford Internment Camp (I)
White Sulphur Springs, WV
Army Service Forces
  1514th Service Command Unit: Ashford Internment Camp (ASF)

Ashford General Hospital POW Camp
Ashford General Hospital, WV
Army Service Forces
  412th Military Police Escort Guard Company (ASF)
  486th Military Police Escort Guard Company (ASF)
  1514th Service Command Unit: Prisoner of War Camp
  3590th Service Command Unit: Prisoner of War Camp
Rainelle Branch POW Camp (I Sub)
  3590th Service Command Unit: Prisoner of War Camp
  3590th Service Command Unit: Detachment, Prisoner of War Camp
Newton D. Baker General Hospital
  3590th Service Command Unit: Prisoner of War Camp
  3590th Service Command Unit: Detachment, Prisoner of War Camp

Acknowledgements
for Assistance with Research

No one writes a book, particularly a non-fiction effort, that doesn't require the help of a great many people. The longer the development period, the longer the finished volume and the greater the amount of detail to authenticate one's work, the greater reliance upon knowledgeable people to bring their special expertise into focus in behalf of the research effort.

This project came about in an unorthodox manner. In the development of a novel about two World War II figures, to lend the highest possible degree of authenticity to the work, the author continued to see a reference book that provided the information he sought. Every possible on-line source was searched as well as libraries, Barnes & Noble, Borders and more. The book simply did not, and does not exist. By default, then, the author finished the novel and began immediately to document the material contained herein.

As stated at the beginning, a detailed endeavor of this nature takes a great deal of talented help to be able to pull together as well as sustained moral encouragement. This element, the moral encouragement was provided in abundance by my partner in the research effort. Additionally, we had to take advantage of numerous libraries and research facilities around the country and tapped into countless websites. The latter have been cited as footnotes and a note of gratitude for their placement on the internet is acknowledged here. However, there were numerous individuals who must be cited - inevitably, some will be omitted because their names are buried somewhere in the mountain of paper and materials collected during the 10 years that has brought us to this point.

In the states that comprised the Army's **Fifth Service Command** (during WWII), the following individuals are acknowledged for their contributions:

Source	**Role**	**Help with -**
James Tobias	U.S. Army Center for Military History	Station List of the Army of the U.S.
Stephen Everett	U.S. Army Center for Military HIstory	Military terminology
Richard E. Osborne	(independent military writer)	General reference *
Douglas Joest	Finance Manager, Evansville Airport	Evansville Municipal Airport
(unnamed)	Librarian, Windfall, Indiana	Windfall POW Camp
Ashley M. Fowler	Museum Liaison	Camp Atterbury
Jim R. Osborne	Indiana Military Museum, Director	General Reference
Wendell Ross	Atterbury-Bakalar Air Museum	Atterbury Army Air Field
James Sellars, Jr.	Atterbury-Bakalar Air Museum	Atterbury Army Air Field
Gordon Lake	Atterbury-Bakalar Air Museum	Atterbury Army Air Field
Kathy Kirkpatrick	POW Camps *(throughout the U.S.)*	Provost Marshal General Lists **

* Through his book, World War Sites in the United States - A Tour Guide & Directory.

** Ms. Kirkpatrick conducts extensive genealogical research with an emphasis on Italian POWs. She was kind enough to provide us with a complete set of the Provost Marshal's digitized files showing each camp's operations and prisoner counts.

U.S. Army Installations - World War II: A Research Reference

Terminology

A fort, is a fort, is a fort. Not really.

There was a unique typology of military installations types, all of which are simply called *stations* by those who kept the official records during World War II. Our work included the "forts" but also included air fields, general hospitals, military academies, depots, ordnance plants and works and much more. It is the word, forts, however that became an issue to define and the following terms evolved as a consequence.

Fort — the term has a fair degree of flex within it in terms of how the word is applied to military installations. A fort is a military facility used for a wide variety of specific applications, e.g., defense, training, administration, troop / unit garrisons. The term implies a permanency to the facility as opposed to the less-than-permanent definition associated with a camp. Yet, numerous forts have been closed, thus terminating their permanency while numerous camps, seemingly less-than-permanent, have remained either as upgrades to fort status or remain as camps, per se.

Camp — the terms, camp and fort, are largely interchangeable (to the non-military) but each has specific meaning to the Army. Camps were intended to be temporary points to assemble and train troops for the war effort but a number have lasted various closure efforts and a number have been upgraded to permanency, e.g., Camp Campbell became Fort Campbell, Camp McCoy became Fort McCoy, Camp Rucker became Fort Rucker, etc.

Cantonment — an ill-defined term. Initially intended to be a temporary facility, but not all were. Some cantonments were redesignated as forts. The practice of naming posts as cantonments had nothing to do with the presumed permanence of the post. It was a common term in use west of the Mississippi during the 1800s.[1]

Military Reservation — this is land set aside for Army purposes and includes grounds on which a military post or station is located. Ergo, all posts, all camps, etc. are military reservations, but not all military reservations are posts etc.

Presidio — by definition, in a military sense, a presidio is either the garrison of a plaza, or fort, or the town or fort which is so garrisoned. In a non-military sense, a presidio is also a prison. These are found in California today and reflect the Spanish influence of earlier times.

Reception Center — a place where recruits and newly inducted personnel are temporarily sent to be examined, classified, equipped, immunized and where basic military principles and regulations are provided before being forwarded to their assigned organizations.

Replacement Depot — miliary establishment, usually located in the Theater of Operations, where replacements are assembled and then distributed to fill vacancies in organizations and / or units. Certain stations in the Zone of the Interior (U.S.) were also replacement depots.

---

[1] An interesting side note. In the conduct of the research for this series, the author found a very common mispronounciation of the term cantonment. Many pronounced the word with an emphasis on the "ton" calling it "tone." The dictionary pronounces the word an kan-ton-muhnt with the "ton" sounding like the measure of weight.

U.S. Army Installations - World War II: Fifth Service Command

Bibliography

Army Ground Forces
Forty, George, **U.S. Army Handbook 1939-1945**, New York, Barnes & Noble, 1998.

Gawne, Jonathan, **Finding Your Father's War**, Philadelphia, Casement Publishers, 2006.

Wilson, John B., **Maneuvers and Firepower: The Evolution of Divisions and Separate Brigades**, Washington, D.C., Center for Military History, 1998.

Stanton, Shelby L., **Order of Battle: U.S. Army - World War II**, Novato, California, 1984.

Prucha, Francis Paul, **Guide tothe Military Posts of the United States - 1789-1895**, The State Historical Society of Wisconsin, 1964.

Cragg, Dan, **Guide to Military Installations**, 6th Edition, Stackpole Books, 2000.

Evinger, William R., **Directory of U.S. Military Bases Worldwide**, Oryx Press, 1995.

Osborne, Richard E., **World War II Sites in the United States: A Tour Guide & Directory,** Riebel-Roque Publishing Company, Indianapolis, IN, 1996.

Office of the Chief of Engineers, **U.S. Army Owned, Sponsored and Leased Facilities, 31 December 1945**, Washington, D.C.

Roberts, Robert B., **Encyclopedia of Historic Forts: The Military, Pioneer, and Trading Posts of the United States**, MacMillan Publishing Company, New York, 1988.

Army Air Forces
Bowman, Martin W., **USAAF: 1939-1945**, Mechanicsburg, Pennsylvania, Stackpole Books, 1997.

Mauer, Mauer, **World War II Combat Squadrons of the United States Air Force - The Official Military Record of Every Active Squadron**, New York, Smithmark Publishers, Inc., 1992.

Mauer, Mauer, ed., **Air Force Combat Units of World War II**, Washington, D.C., Office of Air Force History, 1983.

Prisoner of War Camps
Lewis, George and Mewha, John, **History of Prisoner of War Utilization by the United States Army 1776-1945**. Washington, D.C., Center of Military History, 1982.

Geiger, Jeffrey E., **German Prisoners of War at Camp Cooke, California**, Jefferson, North Carolina, McFarland & Company, Inc., Publishers, , 1996.

Daniels, Roger, **The Decision to Relocate the Japanese Amerians**, Philadelphia, J.B. Lippincott Company, 1975.

Keefer, Louis E., **Italian Prisoners of War in America: 1942-1946: Captives or Allies?**, New YOrk, Praeger, 1992

Bosworth, Allan R., **Americas Concentration Camps**, New York, W.W. Norton, Inc., 1967.

* The listing is in no special order within the categories.

About the Author

Mr. Leonard lived as a young boy at the estate of the Commanding General of Camp Atterbury in central Indiana. Just down the road a very short distance was the POW camp where his mother and he would pick strawberries alongside the German POWs. For a young boy it was an extraordinary opportunity to watch paratroop drops in the field across from the house, tanks roll down the road in front of their gate and "play Army" with actual weapons rather than sticks to which many boys would have to resort.

Although this series of books was initiated in 2002, its magnitude has taken 20 years to complete. It is now time to re-enter each of the books to clean up any shortcomings and remaster each volume with what added information has been found through this *journey*.

Mr. Leonard and his wife lived on the beach in southern California (when they were not traveling about looking up existing and already-closed army forts, camps, POW camps, air bases and the rest of what defines the "stations" that were the U.S. Army Installations. They now live in California's *Wine Country* in Sonoma County.

www.ingramcontent.com/pod-product-compliance
Lightning Source LLC
Chambersburg PA
CBHW040910020526
44116CB00026B/16